Memoirs of the American Mathematical Society
Number 194

A. J. Tromba

On the number of simply connected minimal surfaces spanning a curve

Published by the
AMERICAN MATHEMATICAL SOCIETY
Providence, Rhode Island

VOLUME 12 · ISSUE 2 · NUMBER 194 (first of 3 numbers) · NOVEMBER 1977

ABSTRACT

The classical problem of Plateau is studied from the point of view of global nonlinear analysis. It is shown that minimal surfaces of the topological type of the two disc arise as the zeros of a Fredholm vector field on an infinite dimensional manifold. A framework is developed to prove the finiteness of solutions for an open and dense set of curves and to count the number of such solutions according to sign, although the complete results in this direction are not proved in this paper.

AMS(MOS) subject classifications(1970). Primary 53A10, 65N99, 58E99.

Library of Congress Cataloging in Publication Data CIP

Tromba, Anthony J., 1943 -
On the number of simply connected minimal surfaces spanning a curve.

(Memoirs of the American Mathematical Society ; no. 194)
Bibliography: p.
1. Surface, Minimal. 2. Critical point theory (Mathematical analysis) 3. Boundary value problems. I. Title. II. Series: American Mathematical Society. Memoirs ; no. 194.
QA3.A57 no. 194 [QA644] 510'.8s [516'.362]
ISBN 0-8218-2194-6 77-12180

TABLE OF CONTENTS

INTRODUCTION.

During the last century the problem of Plateau attracted the attention of many mathematicians including that of Riemann, Schwarz, and Weierstrass.

The central question was to prove the existence of a simply connected minimal surface spanning a rectifiable curve $\Gamma \subset \mathbb{R}^3$. The soap film experiments of the Belgian physicist Plateau (1801-1883) made is physically obvious that a solution existed, yet it was not until 1931 that a major breakthrough was made when J. Douglas [5] and T. Rado [26] simultaneously and independently demonstrated the existence of at least one solution.

Since that breakthrough a great deal of research has been directed towards the regularity of solutions (e.g. smoothness of solution up to the boundary [10], [11], [13], [17], [19], [42], existence and types of branch points [17], [19], and analytic extension of solutions across the boundary [15], [43]).

It was known quite early that for plane curves the solution was unique. Rado in 1933 [25] showed that if Γ has a simply covered convex curve as its parallel projection on a plane, then there exists only one minimal surface spanning Γ. Only recently have two other cases been discovered where information on the number of solutions could be obtained. The first, due to Nitsche [18], states that if Γ is analytic and the curvature of Γ is less than 4π then there is a unique minimal surface spanning Γ which is free of branch points.

The second case, due to Tomi [30], states that if Γ is analytic, then the set of surfaces assuming the least possible area is finite. Until now these cases constituted all that was known about the finiteness of the number of solutions or their uniqueness. In this paper we shall show how the global analytical techniques, developed in [8], [9], [32], [33] and [35] can be applied to the Plateau problem to show

(1) If $\mathfrak{J}$ denotes the set of embeddings α of S^1 into $\mathbb{R}^3$ with the property that for the curve $\alpha(S^1)$ every simply connected minimal surface spanning $\alpha(S^1)$ is free from branch points, then there is an open and dense subset $\mathcal{V} \subset \mathfrak{J}$ of embeddings for which the number of minimal surfaces spanning the image is finite.

(2) For any embedding α the minimal surfaces spanning $\alpha(S^1)$ arise as the zeros of a Palais-Smale vector field [35] on a Banach manifold. If $\alpha \in \mathfrak{J}$ this vector field has an integer Euler characteristic which measures the Morse number of minimal surfaces spanning $\alpha(S^1)$. With some restriction, this Euler characteristic depends only on the knot type of α.

(3) Any curve Γ "sufficiently close" to a plane curve has a unique simply connected minimal surface spanning it. This theory has been extended to cover the case of all curves. It is a joint result of R. Böhme and the author that an open dense set $\mathcal{V}$ of all curves has the property that each curve in $\mathcal{V}$ admits only finitely many minimal surfaces of the type of the disc which span it.

We shall give a more precise formulation of these results in the subsequent sections of this paper. However we will first begin with a review of the theory of Euler characteristics of vector fields as developed in [35], and then proceed with our study of Plateau's problem.

ON THE NUMBER OF SIMPLY CONNECTED MINIMAL SURFACES SPANNING A CURVE

by A. J. Tromba*

I. A REVIEW OF THE EULER CHARACTERISTIC OF A PALAIS-SMALE VECTOR FIELD.

In this section we shall follow [35] and the reader is referred to that paper for a more detailed treatment of the theory.

Let M be a C^3 Banach Finsler manifold modelled on a Banach space E which admits an equivalent C^2 norm (and hence C^2 partitions of unity). We shall further suppose that associated to M there is a connection map $K: T^2M \to TM$, e.g. see [7].

If E is a Banach space denote by $\mathcal{L}(E)$ the space of linear continuous operators from E to E; by $C(E)$ the compact linear operators from E to itself; by $\mathcal{L}_C(E)$ the linear operators of the form $I + k$, $k \in C(E)$ and I is the identity map; by $G\mathcal{L}(E)$ the invertible elements of $\mathcal{L}(E)$; and finally by $G\mathcal{L}_C(E)$ the intersection $G\mathcal{L}(E) \cap \mathcal{L}_C(E)$. If E has a Schauder basis it is shown in [8] and [22] that $G\mathcal{L}_C(E)$ has the homotopy type of $G\mathcal{L}(\infty)$ the direct limit of finite dimensional subgroups $G\mathcal{L}(n)$ of $G\mathcal{L}(E)$ and thus in particular $G\mathcal{L}_C(E)$ has two components which we shall designate as $G\mathcal{L}_C^+(E)$ anf $G\mathcal{L}_C^-(E)$ where $G\mathcal{L}_C^+(E)$ is the component of the identity. If E does not have a Schauder basis it is still true that $G\mathcal{L}_C(E)$ has two components.

If $X : M \to TM$ is a C^1 vector field we can define the covariant derivative of X with respect to K at a point $p \in M$ by the formula $\nabla X(p) = K \cdot DX(p)$ where $DX(p) : T_pM \to T_{X(p)}(TM)$ is the Fréchet derivative of X

---

* Research partially supported by National Science Foundation grants GP-39060 and MPS72-05055 A02.

Received by the editor August 28, 1975.

at p. Clearly $\nabla X(p) \in \mathcal{L}(T_pM)$. We say that X is <u>Palais-Smale</u> with respect to the connection K if for each $p \in M$, $\nabla X(p) \in \mathcal{L}_C(T_pM)$. X is Fredholm if $X(p)$ is Fredholm for each $p \in M$. By the index of a Fredholm vector field we mean

$$\text{index } X = \dim \ker \nabla X(p) - \dim \text{coker } \nabla X(p);$$

i.e. the dimension of the kernel minus the dimension of the cokernel. If M is connected this index does not depend on p. If M is not connected the index is constant on the components of M and we shall require it to be the same on all components. It follows easily from the theory of linear Fredholm maps that the index of Palais-Smale vector fields is always zero.

Now suppose that X is a C^2 Palais-Smale and has only a finite number of zeros $p_1, \ldots, p_k$ in the interior M^0 of M. Let $\phi_i : \mathcal{U}_i \to \theta_i$ be coordinate charts about p_i where θ_i is a ball centered at $0 \in E$, $\phi_i(p_i) = 0$ and p_i is the only zero of X in $\mathcal{U}_i$. Let $X^{\phi_i} : \theta_i \to E$ denote the principal parts of X in the coordinate patches ϕ_i. Using the degree theory developed in [7] (for a brief account see [8]) one can define the local degree $\deg(X^{\phi_i}, \theta_i, 0)$. This local degree can be shown to be independent of the choice of the ϕ_i and $\mathcal{U}_i$.

We define the <u>Euler</u> <u>characteristic</u> of X by the formula

(1) $$\chi(X) = \sum_i \deg(X^{\phi_i}, \theta_i, 0)$$

Conditions which assert when this is equal to the Euler characteristic of the underlying manifold M are given in [34].

In the case the zeros of X are non-degenerate we can give a particularly simple interpretation of $\chi(X)$.

DEFINITION. A zero $p \in M^0$ of X is non-degenerate if $\nabla X(p) \in G\mathcal{L}_C(T_pM)$.

Suppose p is such a zero of X. It follows easily that p must be isolated. Define the sign of $\nabla X(p)$ by the rule

$$\operatorname{sgn} X(p) = \begin{cases} +1 & \text{if } \nabla X(p) \in G\mathcal{L}^{+}_{C}(T_pM) \\ -1 & \text{if } \nabla X(p) \in G\mathcal{L}^{-}_{C}(T_pM) \end{cases}$$

From the properties of the local degree one can show that if X has a finite number of non-degenerate zeros $\{p_i\}$ then $\chi(X)$ is given by the formula

(2) $$\chi(X) = \sum_i \operatorname{sgn} \nabla X(p_i)$$

REMARK. In [35] it is shown that one can define the Euler characteristic of a class of vector fields larger than the Palais-Smale fields, which we call Rothe fields. The reader is encouraged to consult this paper.

At the end of this section we shall discuss a transversality result which describes a setting in which non-degenerate zeros arise.

Recall that we are assuming that M admits a Finsler structure (e.g. see [21]) $\| \ \| : TM \to R$.

DEFINITION. A vector field X is proper if whenever $\|X(q_n)\| \to 0$, $\{q_n\}$ a sequence in M, then q_n has a subsequence which converges.

The proof of the following is immediate.

PROPOSITION. If X is proper then the zeros of X form a compact subset of M.

A proper vector field with non-degenerate zeros will have only a finite number of such zeros and so, if the vector field is Palais-Smale, its Euler characteristic is given by formula (2). However, using general position arguments [8] it is also possible to define the Euler characteristic of a proper Palais-Smale field X with zeros in M^0 but in this case one cannot give as simple an interpretation as in formulas (1) and (2).

The Euler characteristic by definition is always an integer. We shall interpret this integer as a measure of the "algebraic" number of zeros of a proper Palais-Smale field X even though the actual "geometric" number may be infinite.

Many objects of geometry and analysis arise as the zeros of Palais-Smale or Fredholm vector fields; for example, geodesics joining two points on a finite dimensional Riemannian manifold, harmonic maps between closed Riemannian manifolds, and simply connected minimal surfaces spanning a smooth curve $\Gamma \subset \mathbb{R}^3$. The Euler characteristic of a vector field X is in fact a globalization to manifolds of the classical Leray-Schauder degree. One of the most important properties of the Leray-Schauder degree in applications to analysis is its invariance under homotopy. We shall now give the global analog of this invariance property.

Let $\tilde{\pi} : M \to [0,1]$ be a smooth fibre bundle over the unit interval with each $\pi^{-1}(t) = M_t$ a smooth C^3 Banach manifold on E. Suppose there is a family of connections K_t with each K_t a connection for M_t.

DEFINITION. Two C^2 smooth Palais-Smale vector fields $X_0 : M_0 \to TM_0$ and $X_1 : M_1 \to TM_1$ are equivalent $(X_0 \sim X_1)$ if there is a C^2 vector field $X : M \to TM$ such that

(i) $X_t = X|M_t : M_t \to TM_t$

(ii) X_t is Palais-Smale with respect to K_t

(iii) the zeros of X are compact in M^0.

THEOREM 1 (Euler-Hopf). If $X_0 \sim X_1$ then $\chi(X_0) = \chi(X_1)$.

A modification of this theorem is the main tool used in XIV to study how the number of simply connected minimal surfaces spanning a curve Γ changes as Γ changes.

We shall now discuss conditions which state when the zeros of a family of vector fields parametrized by some manifold will be generically isolated. The next theorem will be the primary tool in showing that for an open dense set of curves $\Gamma \subset \mathbb{R}^3$ the set of minimal surfaces spanning it are finite in number.

Let $\mathcal{A}$ be a Hilbert manifold, $\pi : \mathcal{M} \to \mathcal{A}$ a C^2 Banach fibre bundle over $\mathcal{A}$ (the fibre $\mathcal{M}_a = \pi^{-1}(a)$ is a C^2 Banach manifold). We shall assume that each fibre $\mathcal{M}_a$ has a Finsler structure $\| \ \|_a : T\mathcal{M}_a \to R$ varying continuously on the vertical subbundle $\xi = \bigcup_{p \in \mathcal{M}} T_p\mathcal{M}_a \subset T\mathcal{M}$ of $T\mathcal{M}$.

Let $X : \mathcal{M} \to T\mathcal{M}$ be a C^1 vector field with the property that

(*) $X|\mathcal{M}_a = X_a$ is a Fredholm vector field over $\mathcal{M}_a (X_a : \mathcal{M}_a \to T\mathcal{M}_a)$, with respect to a family of connections $\{K_a\}$. Thus X is a section of the vertical subbundle ξ of $T\mathcal{M}$.

For each zero $m \in \mathcal{M}_a$ of X the Frechet derivative $X_*(m)$ (which equals the covariant derivative $\nabla X(m)$ at a zero) maps $T_m\mathcal{M}$ into inself. It follows from property (*) that $X_*(m) \in T_m\mathcal{M}_a$.

DEFINITION. We shall say that a vector field $X : \mathcal{M} \to T$ <u>satisfying</u> (*) <u>represents a vertical family of Fredholm vector fields</u> $\{X_a\}_{a \in \mathcal{A}}$.

DEFINITION. A vertical family X is vertically transverse to the zero section of $T\mathcal{M}$ if $X_*(m) : T_m\mathcal{M} \to T_m\mathcal{M}_a$ is surjective for each zero m of X. In this case we write $X \pitchfork_v z(T\mathcal{M})$. Geometrically this means that $X(\mathcal{M})$ is transverse to the zero section of the vertical subbundle ξ of $T\mathcal{M}$.

Recall that a vector field $X: \mathcal{N} \to T\mathcal{N}$, on a Finsler manifold is proper if whenever $\|X(q_n)\| \to 0$ then q_n has a subsequence which converges.

Suppose now that $X : \mathcal{M} \to T\mathcal{M}$ is a vertical family of vector fields, with $\| \ \| : \xi \to R$ a Finsler on the vertical subbundle of $T\mathcal{M}$.

DEFINITION. X is a <u>proper vertical family</u> if whenever $\|X(p_n)\| \to 0$ and $\pi(p_n) \to a$ then p_n has a subsequence which converges.

We conclude this section with the following result whose proof can be found in [33].

THEOREM 2. Suppose $X : \mathcal{M} \to T\mathcal{M}$ is a vertical family of Fredholm vector fields of index zero with $X \pitchfork_v z(T\mathcal{M})$. If X is proper then for an open dense subset $\mathcal{V} \subset \mathcal{A}$ X_a has only non-degenerate zeros for $a \in \mathcal{V}$.

II. ANALYTICAL PRELIMINARIES - THE SOBOLEV SPACES.

In this brief section we review some of the basic results on Sobelev spaces of functions which we shall need in the subsequent sections. Most of the proofs can be found in Calderon [3], Palais [23], or Penot [24].

Let V denote a compact n-dimensional manifold and ξ a C^∞ finite dimensional Riemannian vector bundle over V {in most of our examples will be a product bundle $V \times \mathbb{R}^k$ }. Then for $1 \leq p < \infty$ one can define the Sobelev spaces of p-integrable sections of ξ, $L^p(\xi)$. To define p-integrable one needs a strictly positive Borel measure μ; however if V is compact the space $L^p(\xi)$ can be shown to be independent on the choice of μ. If $V = \mathcal{D}$, the unit disc and $\xi = \mathcal{D} \times R$, the trivial line bundle over $\mathcal{D}$, then $L^p(\xi) = L^p(\mathcal{D})$ the p-integrable functions on $\mathcal{D}$. Let $L^{\bar{p}}(\xi)$, $1/p + 1/\bar{p} = 1$ be the dual of $L^p(\xi)$.

Let $L^p_k(\xi)$ be the space of sections whose r^{th} (covariant) derivatives, $0 \leq r \leq k$, are p-integrable over V. If $\xi = V \times \mathbb{R}^n$ we shall often write $L^p_k(\xi)$ as $L^p_k(V, R^n)$. For $k \geq 0$ $L^p_k(\xi)^0$ denotes the closure of $C^\infty_0(\xi)$ (the C^∞ sections with compact support in the interior of V. $L^{\bar{p}}{}_{-k}(\xi)$ is defined to be the dual of $L^p_k(\xi)^0$ under the bilinear pairing

$$(s_1, s_2) \longmapsto \int_V \langle s_1(x), s_2(x) \rangle_x \, d\mu$$

where μ is some strictly positive measure on V and $\langle , \rangle$ is the Riemannian structure on ξ. Therefore every continuous linear functional $\Phi \in [L^p_k(\xi)^0]^*$ can be represented as a section $s_\Phi \in L^{\bar{p}}{}_{-k}(\xi)$ with

$$\Phi(s) = \int_V \langle s, s_\Phi \rangle \, d\mu \ .$$

The spaces $L^p_k(\xi)$ can also be defined for non-integral values of k using the complex method of interpolation or by means of Bessel potentials.

The following two theorems are known collectively as the Sobolev embedding theorems.

THEOREM 1. Let $1 \le p,q < \infty$ and let k and ℓ be real numbers with $k - \frac{n}{p} \ge \ell - \frac{n}{q}$ and $k \ge \ell$. Then $L_k^p(\xi) \subseteq L^q(\xi)$ and the inclusion map is continuous. If $k - \frac{n}{p} > \ell - \frac{n}{q}$ and $k > \ell$ the inclusion map is a compact linear map.

THEOREM 2. Let $1 \le p < \infty$ and let $k - \frac{n}{p} = \ell + \alpha$ is a non-negative integer and $0 < \alpha < 1$. Then $L_k^p(\xi) \subseteq C^{\ell+\alpha}(\xi)$ and the inclusion map is continuous. Here $C^{\ell+\alpha}(\xi)$ represents the Holder space of C^ℓ sections of ξ whose ℓ^{th} derivatives satisfy a Hölder condition with exponent α.

COROLLARY. If $k - \frac{n}{p} > \ell$ then $L_k^p(\xi) \subseteq C^p(\xi)$ and so if $k > \frac{n}{p}$ there is a continuous inclusion of $L_k^p(\xi)$ into the space of continuous sections.

Now let $\xi = V \times \mathbb{R}^m$, $\tilde{\xi} = V \times \mathbb{R}^{\tilde{m}}$ and $\eta = V \times \mathcal{L}(\mathbb{R}^m, \mathbb{R}^{\tilde{m}})$. If $s_1 \in C^\infty(\eta)$ and $s_2 \in C^\infty(\xi)$ we can "multiply" these two sections to obtain a section $s_1 s_2 \in C^\infty(\tilde{\xi})$. We then have

THEOREM 3. If $1 \le p < \infty$, $k > \frac{n}{p}$ then for $0 \ge j \ge -k$ multiplication extends from $C^\infty(\xi) \oplus C^\infty(\eta) \to C^\infty(\tilde{\xi})$ to a continuous bilinear map of $L_k^p(\xi) \oplus L_j^{\overline{p}}(\eta) \to L_j^p(\tilde{\xi})$.

The next result concerns the restriction of functions to the boundary. Let V be a C^∞ n-manifold with non-empty boundary ∂V. Let ξ be a bundle over V with $\xi|\partial V$ the restriction of ξ to ∂V. Then we have a restriction map $\psi : C^\infty(\xi) \to C^\infty(\xi|\partial V)$ defined in the obvious way.

THEOREM 4. The restriction map ψ extends to a continuous linear map of $L_k^p(\xi) \to L_{k-1/2}^p(\xi|\partial V)$.

REMARK. It is not true that the restriction extends to a map $L_k^p(\xi)$ to $L_k^p(\xi|\partial V)$.

REMARK. The spaces $L_k^2(\xi)$ are Hilbert spaces and are often given a special notation $H^k(\xi)$, a convention we shall adopt in the subsequent sections. Thus, in accordance with our previous custom, if $\xi = V \times \mathbb{R}^m$, $H^k(\xi)$ is written as $H^k(V,\mathbb{R}^m)$.

Our last theorem of the section states how the L_k^p maps "behave" under composition by smooth maps [24].

THEOREM 5. Let $\xi = V \times \mathbb{R}^m$, $\tilde{\xi} = V \times \mathbb{R}^{\tilde{m}}$ and $\Xi : \mathbb{R}^m \to \mathbb{R}^{\tilde{m}}$ a C^{r+k} (H^{r+k}) smooth map. Then Ξ if $k > \frac{n}{p}$ induces a C^r map $\Xi_\# : L_k^p(\xi) \to L_k^p(\tilde{\xi})$ by the composition rule

$$\Xi_\#(s)(x) = \Xi(s(x)) .$$

We are now ready to proceed with a global formulation of the Plateau problem.

III. THE GLOBAL FORMULATION OF THE PROBLEM OF PLATEAU.

In this section we show how simply connected minimal surfaces arise as the critical points of a functional on an infinite dimensional manifold of maps.

Let $H^{r+2}(S^1, \mathbb{R}^3)$ be the Sobolev-Hilbert space of H^{r+2} maps of the unit circle S^1 into $\mathbb{R}^3$ with $r \geq 5$. By an easy case of the Sobolev-imbedding theorems (Theorem 2, Section II) $H^{r+2}(S^1, \mathbb{R}^3) \subset C^6(S^1, \mathbb{R}^3)$ the six times continuously differentiable maps.

Let $\mathcal{A} = \text{Emb}(S^1, \mathbb{R}^3)$ be the open submanifold of $H^{r+2}(S^1, \mathbb{R}^3)$ which consists of embeddings of S^1 into $\mathbb{R}^3$. Let Γ be the image of such an embedding $\alpha \in \mathcal{A}$. Let $p \geq 2$ and let $L_2^p(S^1, \Gamma)$ be the Banach submanifold of $L_2^p(S^1, \mathbb{R}^3)$ consisting of those maps which take S^1 to Γ. In [24] it is shown that $L_2^2(S^1, \Gamma) = H^2(S^1, \Gamma)$ is a C^r submanifold of $L_2^2(S^1, \mathbb{R}^3) = H^2(S^1, \mathbb{R}^3)$. Let $\mathcal{N}_\alpha$ be the component of $H^2(S^1, \Gamma)$ determined by the embedding α. Recall that the tangent space to $\mathcal{N}_\alpha$ at a point $u \in \mathcal{N}_\alpha$ can be identified with the H^2 maps h of S^1 into $\mathbb{R}^3$ with $h(\theta) \in T_{u(\theta)}\Gamma$.

Let $\mathcal{M}_\alpha$ be the open submanifold of $\mathcal{N}_\alpha$ consisting of diffeomorphisms.

We can extend every $u \in H^2(S^1, \mathbb{R}^3)$, $u = (u_1, u_2, u_3)$ uniquely to a harmonic function $\hat{u} : \mathcal{D} \to \mathbb{R}^3$. Define the Dirichlet energy functional $E : H^2(S^1, \mathbb{R}^3) \to R$ by

(1)
$$E(u) = \frac{1}{2} \sum_{j=1}^{3} \int_{\mathcal{D}} \left[\left(\frac{\partial \hat{u}_j}{\partial x}\right)^2 + \left(\frac{\partial \hat{u}_j}{\partial y}\right)^2\right] dxdy .$$

Using Green's theorem and integrating by parts we can write E as an integral along the boundary S^1 of $\mathcal{D}$

(2)
$$E(u) = \frac{1}{2} \int_{S^1} \left\langle \frac{\partial u}{\partial r}, u \right\rangle d\theta$$

where u is viewed as a function $u(r,\theta)$ of the polar coordinates (r,θ) for the disc $\mathcal{D}$; $\frac{\partial u}{\partial r}$ is the "normal" derivative of $\hat{u}$ on S^1 ; and $\langle \cdot , \rangle$ is the $\mathbb{R}^3$ inner product.

Since $u : S^1 \to \mathbb{R}^3$ we can represent each component function u_j as a Fourier series $u_j(\theta) = u_j(1,\theta) = \sum_\nu c_\nu^j e^{i\nu\theta}$ with $c_{-\nu}^k = \overline{c_\nu^k}$, the complex conjugate. The harmonic extension of u is given by the formula

$$u_j(r,\theta) = c_\nu^j r^{|\nu|} e^{i\nu\theta}$$

and therefore $\frac{\partial u_j}{\partial r}(\theta) = \frac{\partial u_j}{\partial r}(1,\theta) = \Sigma|\nu|c_\nu^j e^{i\nu\theta}$.

In Fourier series notation (2) transforms to

$$\text{(3)} \qquad E(u) = 2\pi \sum_j \{\sum_\nu |\nu| |c_\nu^j|^2\}$$

It is often times convenient to use the Fourier represrntation

$$u_j(\theta) = \frac{1}{2} a_0^j + \sum_{\nu > 0} (a_\nu^j \cos \nu\theta + b_\nu^j \sin \nu\theta)$$

$$c_\nu^j = \frac{1}{2} (a_\nu^j - ib_\nu^j), \; c_{-\nu}^j = \frac{1}{2}(a_\nu^j + ib_\nu^j).$$

With this representation (3) becomes

$$\text{(3')} \qquad E(u) = \pi \sum_{j=1}^{3} \left\{ \sum_{\nu=1}^{\infty} \nu[(a_\nu^j)^2 + (b_\nu^j)^2] \right\} .$$

Since $H^2(S^1, \mathbb{R}^3) \subset H^1(S^1, \mathbb{R}^3) = \{u : S^1 \to \mathbb{R}^3 \,|\, \forall_j , \sum_\nu |\nu|^2 \,|c_\nu^j|^2 < \infty\}$ it follows that E is continuous and quadratic on H^2 and hence a C^∞ map on $H^2(S^1, \mathbb{R}^3)$. Therefore the restriction E_α of E to $\mathcal{N}_\alpha$ is a C^r smooth map. Let $\overline{\mathcal{M}}_\alpha$ be the closure of $\mathcal{M}_\alpha$ in $\mathcal{N}_\alpha$. In this section we shall show that the critical points of E_α in $\overline{\mathcal{M}}_\alpha$ are the simply connected minimal surfaces spanning Γ.

In order to do this we must review the classical definition of simply connected minimal surface. For more details on the original approach of Douglas and Rado to this problem see [5], [25], and for a more elegant approach see Courant [4].

DEFINITION. Let Γ be a rectifiable simple closed curve in $\mathbb{R}^3$. A simply connected minimal surfaces spanning Γ is a map $u : \mathcal{D} \to \mathbb{R}^3$, $\mathcal{D}$ the unit disc in $\mathbb{R}^2$, which satisfies

(i) $\Delta u = \dfrac{\partial^2 u}{\partial x^2} + \dfrac{\partial^2 u}{\partial y^2} = 0$, on the interior of $\mathcal{D}$.

Let (r,θ) denote the polar coordinates of $\mathcal{D}$. Then in polar coordinates (i) becomes

(i') $\dfrac{\partial^2 u}{\partial r^2} + \dfrac{1}{r^2}\dfrac{\partial^2 u}{\partial \theta^2} + \dfrac{1}{r}\dfrac{\partial u}{\partial r} = 0$

(ii) $\langle u_r, u_\theta \rangle = u_r \cdot u_\theta = 0$ or in rectangular coordinates $\langle u_x, u_y \rangle = 0$

(iii) $\|u_r\|^2 = (1/r^2)\,\|u_\theta\|^2$ or in rectangular coordinates $\|u_x\|^2 = \|u_y\|^2$

where $\|\ \| = \sqrt{\langle\, ,\, \rangle}$ denotes the Euclidean norm on $\mathbb{R}^3$.

(iv) $u|S^1 = u|\partial\mathcal{D}$ takes S^1 homeomorphically onto Γ.

WARNING: In the remainder of this paper we shall use simply connected minimal surface and minimal surface interchangeably.

THEOREM 1. The critical points of E_α in $\mathcal{N}_\alpha$ satisfy (i), (ii) and (iii) above.

Proof. Condition (i) is satisfied automatically. Using representation (2) for $E_\alpha : \mathcal{N}_\alpha \to R$ an easy and direct computation shows that the Frechet derivative $DE_\alpha(u) : T_u\mathcal{N}_\alpha \to R$ is given by

(4) $$DE_\alpha(u)[h] = \int_{S^1} \left\langle \frac{\partial u}{\partial r}, h \right\rangle d\theta .$$

Let $\mathcal{OP}(\mathbb{R}^3)$ represent the orthogonal projections on $\mathbb{R}^3$ and let $\Omega : \Gamma \to \mathcal{OP}(\mathbb{R}^3)$ be defined by setting $\Omega(p)$, $p \in \Gamma$, equal to the orthogonal projection of $\mathbb{R}^3$ onto $T_p\Gamma$, the tangent space to Γ at p.

It is a standard result that Ω is C^{r+1} smooth, and also that the composition $u \longmapsto \Omega \circ u$ is C^{r-1} smooth.

Continuing equation (4) we have

$$DE_\alpha(u)[h] = \int_{S^1} \left\langle \Omega(u(\theta)) \frac{\partial u}{\partial r}(\theta), h(\theta) \right\rangle d\theta . \tag{5}$$

For all $\theta \in [0, 2\pi]$, $\Omega(u(\theta)) \frac{\partial u}{\partial r}(\theta) \in T_{u(\theta)}\Gamma$. If u is a critical point of (5) $DE_\alpha(u)[h] = 0$ for all $h \in H^2(S^1, \mathbb{R}^3)$ such that $h(\theta) \in T_{u(\theta)}\Gamma$. This implies that for all θ, $\Omega(u(\theta)) \frac{\partial u}{\partial r}(\theta) = 0$. Since for $r = 1$ $\frac{\partial u}{\partial \theta}(\theta) \in T_{u(\theta)}\Gamma$ we have that

$$\frac{\partial u}{\partial \theta} \cdot \frac{\partial u}{\partial r} = 0 \quad \text{for } r = 1 \text{ and for all } \theta . \tag{6}$$

We shall show that the harmonic property of u together with (6) implies properties (ii) and (iii). Equations (ii) and (iii) are known as the conformality conditions and they can be expressed quite neatly using complex notation. Let $\tilde{u}$ be the vector harmonic conjugate of $u = (u_1, u_2, u_3)$. Let $f(z) = (f_1, f_2, f_3) = u + i\tilde{u}$. Then it is easy to check that (ii) and (iii) are equivalent to

$$[f'(z)]^2 = \sum_{i=1}^{3} [f_j'(z)]^2 \equiv 0$$

Now consider $u_r - i u_\theta = u_r + i \tilde{u}_r$. On S^1 $u_r - i u_\theta = z \cdot f'(z)$ and $[u_r - i u_\theta]^2 = \|u\|^2 - \|u_\theta\|^2 - 2i \langle u_r, u_\theta \rangle$ where $[u_r - i u_\theta]^2 = \sum_{j=1}^{3} [u_r^j - i u_\theta^j]^2$. Therefore $\Omega(u) \frac{\partial u}{\partial \theta} = 0$ on S^1 implies that the imaginary part of $z^2[f'(z)]^2$ is zero on S^1 which further implies that $z^2[f'(z)]^2$ is a constant. But this vanishes at zero and hence $[f'(z)]^2 \equiv 0$. This concludes the proof of Theorem 1.

The following theorem is a standard result in the theory of minimal surfaces. Its proof can be found in [4].

THEOREM 2. Let Γ be a rectifiable simple closed curve in $\mathbb{R}^3$. Suppose u is a map from S^1 to Γ which maps S^1 monotonically onto Γ and satisfies (i),(ii), and (iii). Then u cannot be constant on any interval of S^1.

THEOREM 3. If $u \in \overline{\mathcal{M}}_\alpha \subset \mathcal{N}_\alpha$ is a critical point of E_α then u satisfies (i), (ii), and (iv).

Proof. By Theorem 1 u satisfies (i),(ii), and (iii). Since $u \in \overline{\mathcal{M}}_\alpha$, u is monotonic and by Theorem 2 cannot be constant on any interval. Thus u is locally one to one. But u is homotopic to an embedding and is monotonic and therefore it must globally one to one and hence a homeomorphism.

Putting the results of Theorems 1 and 3 together we obtain the main result of this section.

THEOREM 4. The critical points of $E_\alpha : \mathcal{N}_\alpha \to R$ in $\overline{\mathcal{M}}_\alpha$ are the classical simply connected minimal surfaces spanning the curve $\Gamma = \alpha(S^1)$.

It was in 1931 that both J. Douglas and T. Rado first proved the existence of a critical point of E in $\overline{\mathcal{M}}_\alpha$. For this work J. Douglas, in 1936, received one of the two first (along with Lars Ahlfors) Fields Medals in mathematics.

It would be valuable if a modern proof of existence could be given along the lines of the Morse theory as developed by Palais and Smale. For the last few yars th principal global analytical tool to demonstrate the existence of critical points in the variational calculus has been the condition (C) of Palais and Smale. However for condition (C) to hold the functional must determine the space of mappings one uses. If this space is outside the Sobolev range, you are dead-ended, and the manifold of maps approach won't help much.

The space $\mathcal{N}_\alpha$ is a Hilbert manifold with a natural H^2 Riemannian structure induced from the Riemannina structure on $\mathbb{R}^n$. Thus one can take the gradient of E_α with respect to this structure. It follows from results in [32] that the functional E cannot satisfy condition (C). In order to force condition (C) to apply to this situation one would have to take the Sobelev space of maps $H^{1/2}(S^1, \mathbb{R}^3)$ on which E is "naturally" defined. Recall that the $H^{1/2}$ norm of a map $u : S^1 \to \mathbb{R}^3$ is given by $\|u\|^2_{H^{1/2}} = \int_{S^1} \langle u,u \rangle d\theta + \int_{S^1} \langle \frac{\partial u}{\partial r}, u \rangle d\theta$ the second term of which is the functional E. But since $H^{1/2}$ "just misses" being in the Sobelev range (i.e. $H^{1/2}$ is not contained in the space of continuous functions on S^1) we cannot even define the manifold $H^{1/2}(S^1,\Gamma)$ and so this approach breaks down completely.

In [32] and [37] a new approach to variational theory has been put forth which eliminates the need for condition (C) and shows that the choice of space for your variational problem is essentially irrelevant as long as it is in Sobelev range. The main thrust of this point of view is to switch attention from a functional $f : M \to R$ on a manifold M to an associated vector field X with $X(f) \geq 0$ and whose zeros are precisely the critical points of f. In the next section we shall prove the existence of such a vector field for the Dirichlet functional $E : \mathcal{N}_\alpha \to R$.

IV. THE EXISTENCE OF A VECTOR FIELD ASSOCIATED TO THE DIRICHLET FUNCTIONAL E_α .

We say that a vector field X on a manifold M is associated to a functional $f: M \to R$ if the critical points of f are the zeros of X. In this section we shall prove the existence of a vector field X^α on $\mathcal{N}_\alpha$ which is associated to E_α and also transverse $(X^\alpha(E_\alpha) \geq 0)$ to E_α. From Theorem 4 of the previous section it follows that the zeros of X^α in $\overline{\mathcal{M}}_\alpha$ are the classical simply connected minimal surfaces spanning $\Gamma = \alpha(S^1)$. In the next section we shall show that X^α is Palais-Smale with respect to a specific connection.

In order to carry this program out we will need to employ some of the more exotic Sobelev spaces of maps. The two principal spaces we shall use are $H^{1/2}(S^1,\mathbb{R}^3)$ and $H^{-1/2}(S^1,\mathbb{R}^3)$. The space $H^{1/2}$ was mentioned in Section II and formally defined at the end of the last section. The space $H^{-1/2}$ is the dual of $H^{1/2}$ under the bilinear pairing $(f,g) = \int_{S^1} \langle f,g\rangle\, d\theta$.

An easy Fourier series characterization of the space $H^{1/2}(S^1,R)$ and $H^{-1/2}(S^1,R)$ can be given. The space $H^{1/2}(S^1,R)$ can be thought of as those real-valued functions $f: S^1 \to R$ defined almost everywhere and whose Fourier series $f(\theta) \sim a_0/2 + \sum_\nu (a_\nu \cos \nu\theta + ib_\nu \sin \nu\theta)$ satisfies $\sum \nu(a_\nu^2 + b_\nu^2) < \infty$. The space $H^{-1/2}(S^1,R)$ consists of those f whose Fourier series satisfies $\sum \frac{1}{\nu}(a^2 + b^2) < \infty$.

Let $H^1(S^1,\Gamma)$ be the H^1 Hilbert manifold of H^1 maps of S^1 to Γ. For $u \in H^1(S^1,\Gamma)$ define the Hilbert space $H^{1/2}(S^1,T\Gamma)_u$ to be the $H^{1/2}(S^1,\mathbb{R}^3)$ completion of those functions $h \in H^1(S^1,\mathbb{R}^3)$ with $h(\theta) \in T_{u(\theta)}\Gamma$ for all θ . For brevity let us denote the space $H^{1/2}(S^1,T\Gamma)_u$ by $H_u^{1/2}$.

THEOREM 1. The space $H_u^{1/2}$ can be characterized as the subspace of $H^{1/2}(S^1,\mathbb{R}^3)$ with $h(\theta) \in T_{u(\theta)}\Gamma$ for almost all $\theta \in S^1$.

Proof. Let $\Omega\colon \Gamma \to \mathcal{O}\mathcal{P}(\mathbb{R}^3)$ be C^5 the map introduced in Section III. From Theorem 5 Section II it follows that the composition map $u(\theta) \to \Omega(u(\theta))$ is C^4 smooth and the map $S^1 \to \mathcal{L}(\mathbb{R}^3,\mathbb{R}^3)$ given by $\theta \to \Omega(u(\theta))$ is of class H^2.

If $h \in H^{1/2}(S^1,\mathbb{R}^3)$ and $h(\theta) \in T_{u(\theta)}\Gamma$ for almost all θ it follows that $\Omega(u(\theta))h(\theta) = h(\theta)$ for almost all θ. Let $\{g_n\}$ be a sequence in $H^1(S^1,\mathbb{R}^3)$ converging to h in $H^{1/2}$. By Theorem 3 of Section II $h_n(\theta) = \Omega(u(\theta))g_n(\theta)$ is a sequence in H^1 with $h_n(\theta) \in T_{u(\theta)}\Gamma$ and which converges (by the same result) to $\Omega(u(\theta))h(\theta) = h(\theta)$ almost everywhere. We have proben Theorem 1 in one direction.

Conversely, suppose $h \in H_u^{1/2}$. Then h is the $H^{1/2}$ limit of functions $\{h_n\}$ in $H^1(S^1,\mathbb{R}^3)$ with $h_u(\theta) \in T_{u(\theta)}\Gamma$ for all θ. Thus

(1) $$\Omega(u(\theta))h_n(\theta) = h_n(\theta)$$

for all θ. By Theorem 3 Section II the left hand side tends to $\Omega(u(\theta))h(\theta)$ in $H^{1/2}$ norm and the right hand side by definition tends to $h(\theta)$ in $H^{1/2}$ norm. Since $\Omega(u)h$ and h are both in L_2 and equal it follows that $\Omega(u(\theta))h(\theta) = h(\theta)$ for almost all θ and the proof of Theorem 1 is complete.

For fixed $u \in H^2(S^1,\Gamma)$ define the functional $\Phi_u\colon H_u^{1/2} \to R$ by

$$\Phi_u(\beta) = \frac{1}{2}\int_{S^1}\left\langle \frac{\partial u}{\partial r} - \frac{\partial \beta}{\partial r}, u - \beta \right\rangle d\theta \; ^{*} .$$

---

*Recall that by $\frac{\partial \beta}{\partial r}(\theta)$, $\frac{\partial u}{\partial r}(\theta)$ we mean the normal derivatives at θ of the harmonic extension of β and u to the unit disc.

Suppose $u^j(\theta) = \frac{1}{2} a_0^j + \sum \left(a_\nu^j \cos \nu\theta + i\, b_\nu^j \sin \nu\theta\right)$

$$\beta^j(\theta) = \frac{1}{2} d_0^j + \sum \left(d_\nu^j \cos \nu\theta + i\, e_\nu^j \sin \nu\theta\right) .$$

Then in Fourier series representation

$$\Phi_u(\beta) = \pi \sum_{j=1}^{3} \sum_{\nu=1}^{\infty} \nu[(a_\nu^j - d_\nu^j)^2 + (b_\nu^j - e_\nu^j)^2] .$$

Clearly $\Phi_u(\beta) \geq 0$ for all $\beta \in H_u^{1/2}$.

THEOREM 2. The functional Φ_u assumes a unique minimum at a point $X(u) \in H_u^{1/2}$.

Proof. Set $\rho = \inf_\beta \Phi_u(\beta)$ and let $\beta_n \in H_u^{1/2}$ be a sequence with $\Phi_u(\beta_n) \to \rho$. We would like to show that $\{\beta_n\}$ is bounded in $H_u^{1/2}$. This follows immediately from the following.

LEMMA. The norm $|||\ \ |||$ on $H_u^{1/2}$ defined by

$$|||h|||^2 = \int_{S^1} \left\langle \frac{\partial h}{\partial r}, h \right\rangle d\theta$$

is equivalent to the $H^{1/2}$ norm on $H_u^{1/2}$ defined by

$$\|h\|^2_{H^{1/2}} = \int_{S^1} \langle h,h \rangle d\theta + \int_{S^1} \left\langle \frac{\partial h}{\partial r}, h \right\rangle d\theta .$$

Proof (of lemma). Clearly

(2) $$|||\ \ ||| \leq \|\ \ \|_{H^{1/2}} .$$

We must check that $|||\ \ |||$ is in fact a norm; that is, if $|||h||| = 0$ then $h = 0$. But if $|||h||| = 0$ then all but the first Fourier coefficients of the component functions of h vanish. Therefore h must be a constant function. By Theorem 1 $h(\theta) \in T_{u(\theta)}\Gamma$ for almost all θ , and since h is continuous, this must hold for all θ. Let θ_1 and θ_2 be two values of θ where $T_{u(\theta_0)}\Gamma \neq T_{u(\theta_1)}\Gamma$. Now $h(\theta_s) \in T_{u(\theta_s)}\Gamma$, $s = 0,1$. Since each $T_{u(\theta_s)}\Gamma$ is one dimensional and h is constant if $h \neq 0$, then this would imply that $T_{u(\theta_0)}\Gamma = T_{u(\theta_1)}\Gamma$ a clear contradiction. Thus $h = 0$ and $|||\ |||$ is a norm on $H_u^{1/2}$. It is not hard to see that $|||\ |||$ is a complete norm for $H_u^{1/2}$. Using inequality (2) and applying the open mapping theorem we see that the $H^{1/2}$ norm and $|||\ \ |||$ are indeed equivalent.

Thus the sequence $\{\beta_n\}$ is a bounded sequence in the norm $|||\ \ |||$ and hence a bounded sequence in the $H^{1/2}$ norm. Consequently $\{\beta_n\}$ has a weakly convergent subsequence $\{\beta_{n_j}\}$ which converges to an element $X^\alpha(u)$. One checks easily that Φ_u is lower-semi-continuous with respect to weak convergence and so $\Phi_u(X^\alpha(u)) = \rho$. We must show that $X^\alpha(u)$ is unique.

The map which sends $\beta \to \Phi_u(\beta)$ is clearly differentiable and has a Fréchet derivative at X^α which is a linear map from $H_u^{1/2}$ to the reals given by

$$D\Phi_u(X^\alpha)[h] = \int_{S^1} \left\langle -\frac{\partial u}{\partial r} + \frac{\partial X^\alpha}{\partial r}, h \right\rangle d\theta .$$

Since X is a minimum this derivative vanishes and so we find that

$$\int_{S^1} \left\langle \frac{\partial u}{\partial r}, h \right\rangle d\theta = \int_{S^1} \left\langle \frac{\partial X^\alpha}{\partial r}, h \right\rangle d\theta \tag{3}$$

for all $h \in H_u^{1/2}$.

Suppose that the minimum is assumed at two points X_1^α and X_2^α. From (3) it follows that $\int_{S^1} \left\langle \frac{\partial X_1^\alpha}{\partial r}, h \right\rangle d\theta = \int_{S^1} \left\langle \frac{\partial X_2^\alpha}{\partial r}, h \right\rangle d\theta$ for all $h \in H_u^{1/2}$. This implies that $|||X_1^\alpha - X_2^\alpha||| = 0$ and by the lemma we see that $X_1^\alpha = X_2^\alpha$.

THEOREM 3. The vector field $X^\alpha(u)$ over u, $X^\alpha(u) \in H_u^{1/2}$, satisfies the relation

$$\Omega(u)\,\frac{\partial X^\alpha}{\partial r}(u) = \Omega(u)\,\frac{\partial u}{\partial r}$$

where $u \longmapsto \Omega(u)$ is the C^{r-1} composition of the map $\Omega\colon \Gamma \to \mathcal{O}\mathcal{P}(R^3)$ and $u \in H^2(S^1,\Gamma)$.

<u>Proof</u>. If $X^\alpha(u) \in H^{1/2}(S^1,\mathbb{R}^3)$ then $\frac{\partial X^\alpha}{\partial r}(u)$ $H^{-1/2}(S^1,\mathbb{R}^3)$. By Theorems 3 and 5 of Section II $\Omega(u)\,\frac{\partial X}{\partial r}(u) \in H^{-1/2}$. In the proof of Theorem 2 we showed that $\int_{S^1}\left\langle \frac{\partial X^\alpha}{\partial r}(u),h\right\rangle d\theta = \int_{S^1}\left\langle \frac{\partial u}{\partial r},h\right\rangle d\theta$ for all $h \in H_u^{1/2}$. Since $\Omega(u)[h] = h$ for all $h \in H_u^{1/2}$ we see that

$$\int_{S^1}\left\langle \Omega(u)\,\frac{\partial X^\alpha}{\partial r}(u),h\right\rangle d\theta = \int_{S^1}\left\langle \Omega(u)\,\frac{\partial u}{\partial r},h\right\rangle d\theta$$

which implies that for almost all θ, $\Omega(u(\theta))\,\frac{\partial X^\alpha}{\partial r}(u)(\theta) = \Omega(u(\theta))\,\frac{\partial u}{\partial r}(\theta)$ which is what we wanted to establish.

Our first objective now will be to show that $X^\alpha(u) \in H_u^{1/2} \subset H^{1/2}(S^1,\mathbb{R}^3)$ is in addition an element of $H^2(S^1,\mathbb{R}^3)$. In order to do this we shall have to appeal to some standard regularity results in the theory of elliptic partial differential equations.

The map $\Omega(u) \in H^2(S^1,\mathcal{L}(\mathbb{R}^3,\mathbb{R}^3))$ can be considered as a 3×3 matrix whose entries $a_{jk}(\theta) \in H^2(S^1,\mathbb{R})$. If $f \in H^{1/2}(S^1,\mathbb{R})$ denote by $\hat{f}$ its unique harmonic extension to the unit disc. By $\hat{\Omega}(u)$ we mean the C^∞ map of the unit disc $\mathcal{D}$ onto $\mathcal{L}(\mathbb{R}^3,\mathbb{R}^3)$ whose jk^{th} entry is $\hat{a}_{jk}\colon \mathcal{D} \to R$. Thus $\hat{\Omega}(u)$ is a function of the polar coordinates (r,θ), $0 \le r \le 1$, $0 \le \theta \le 2\pi$ with the Laplacean $\Delta\hat{\Omega}(u) = 0$ and $\hat{\Omega}(u)(1,\theta) = \Omega(u(\theta))$. If $h \in H^{1/2}(S^1,\mathbb{R}^3)$, let $\hat{h}\colon \mathcal{D} \to \mathbb{R}^3$ denote its harmonic extension by components.

On S^1, $\Omega(u(\theta))[X^\alpha(u)(\theta)] = \hat{X}^\alpha(u)(\theta)$. Therefore $\widehat{\Omega(u)[X^\alpha(u)]} = \hat{X}^\alpha(u)$. By how much does $\hat{\Omega}(u)\hat{X}^\alpha(u)]$ differ from $\widehat{\Omega(u)[X^\alpha(u)]}$? This is answered in the following

THEOREM 4.

$$\hat{\Omega}(u)[\hat{X}^\alpha(u)] = \hat{X}^\alpha(u) + \psi \tag{4}$$

where $\psi \in H^2(\mathcal{D},\mathbb{R}^3)^0$.

Proof. If $f\colon \mathcal{D} \to R$, define $\nabla f\colon \mathcal{D} \to \mathbb{R}^2$ by $\nabla f = \left(\frac{\partial f}{\partial x}, \frac{\partial f}{\partial y}\right)$

Formally

$$\Delta(fg) = (\Delta f)g + (\Delta g)f + 2\langle \nabla f, \nabla g\rangle_{\mathbb{R}^2} \tag{5}$$

where $\langle\ ,\ \rangle_{\mathbb{R}^2}$ is the $\mathbb{R}^2$ dot product. Consequently if f and g are harmonic $\Delta(fg) = 2\langle \nabla f, \nabla g\rangle_{\mathbb{R}^2}$. If $h\colon \mathcal{D} \to \mathbb{R}^3$, $h = (h_1,h_2,h_3)$ define $\nabla\hat{\Omega}(u) \cdot \nabla h\colon \mathcal{D} \to \mathbb{R}^3$ to be the map whose ℓ^{th} row (column; depending on how your write vectors) is $\sum_j \langle \nabla \hat{a}_{\ell j}, \nabla h_j\rangle_{\mathbb{R}^2}$.

Now $\hat{X}^\alpha(u) \in H^1(\mathcal{D},\mathbb{R}^3)$ and $\hat{\Omega}(u) \in H^{5/2}(\mathcal{D},\mathcal{L}(\mathbb{R}^3,\mathbb{R}^3))$. Thus

$$\nabla\hat{\Omega}(u) \cdot \nabla\hat{X}(u) \in H^0(\mathcal{D},\mathbb{R}^3) = L_2(\mathcal{D},\mathbb{R}^3).$$

Let ψ be the unique solution in $H^2(\mathcal{D},\mathbb{R}^3)^0$ to Laplace's equation

$$\Delta\psi = 2\nabla\hat{\Omega}(u) \cdot \nabla\hat{X}^\alpha(u) \quad .^*$$

Then, as in formula (5), we have

$$\Delta\{\hat{\Omega}(u)[\hat{X}^\alpha(u)] - \hat{X}^\alpha(u)\} = 2\nabla\hat{\Omega}(u) \cdot \nabla X^\alpha(u) = \Delta\psi .$$

---

* We are using here a standard regularity result from elliptic P.D.E.

But $\{\hat{\Omega}(u)[\hat{X}^\alpha(u)] - \hat{X}^\alpha(u)\}|S^1 = 0$ almost everywhere, and so $\{\hat{\Omega}(u)[\hat{X}^\alpha(u)] - \hat{X}^\alpha(u)\} \in H^2(\mathcal{D}, \mathbb{R}^3)^0$ which implies that

(6) $$\psi = \hat{\Omega}(u)[\hat{X}^\alpha(u)] - \hat{X}^\alpha(u) \quad .$$

We next prove

THEOREM 5. Let $X^\alpha(u)$ be the unique vector field over u in $H^{1/2}(S^1, \mathbb{R}^3)$ satisfying

$$\Omega(u) \frac{\partial X^\alpha}{\partial r} = \Omega(u) \frac{\partial u}{\partial r} \quad .$$

Then $X^\alpha(u) \in H^2(S^1, \mathbb{R}^3)$.

<u>Proof.</u> Differentiating (6) with respect to r and setting $r = 1$ we see that

$$\frac{\partial \hat{\Omega}}{\partial r}(u)[X^\alpha(u)] + \Omega(u) \frac{\partial X^\alpha}{\partial r}(u) - \frac{\partial X^\alpha}{\partial r}(u) = \frac{\partial \psi}{\partial r}$$

$$= \frac{\partial \hat{\Omega}}{\partial r}(u)[X^\alpha(u)] + \Omega(u) \frac{\partial u}{\partial r} - \frac{\partial X^\alpha}{\partial r}(u) = \frac{\partial \psi}{\partial r} \quad .$$

Thus

(7) $$\frac{\partial X^\alpha}{\partial r}(u) = \frac{\partial \hat{\Omega}}{\partial r}(u)[X^\alpha(u)] + \Omega(u) \frac{\partial u}{\partial r} - \frac{\partial \psi}{\partial r} \quad .$$

We know that

$$\frac{\partial \hat{\Omega}}{\partial r}(u) \in H^1(S^1, \mathcal{L}(\mathbb{R}^3, \mathbb{R}^3)) \quad ,$$

and consequently applying Theorem 3 of Section II we see that $\frac{\partial \hat{\Omega}}{\partial r}(u)[X^\alpha(u)] \in H^{1/2}(S^1, \mathbb{R}^3)$. We also know that $\Omega(u) \frac{\partial u}{\partial r} \in H^1(S^1, \mathbb{R}^3)$, and by applying Theorem 4 of Section II we also have $\frac{\partial \psi}{\partial r} \in H^{1/2}(S^1, \mathbb{R}^3)$.

From the existence proof for $X^\alpha(u)$ we saw that $\frac{\partial X^\alpha}{\partial r}(u) \in H^{-1/2}(S^1,\mathbb{R})^3$ but equation (7) tells us that (surprisingly) $\frac{\partial X^\alpha}{\partial r}(u) \in H^{1/2}(S^1,\mathbb{R}^3)$. Thus $X^\alpha(u) \in H^{3/2}(S^1,\mathbb{R}^3)$, and $\nabla\hat{\Omega}(u) \cdot \nabla\hat{X}^\alpha(u) \in H^1(\mathcal{D},\mathbb{R}^3)$ (this is not completely obvious). This says that the ψ of Theorem 4 belongs to the space $H^3(\mathcal{D},\mathbb{R}^3)^0$, which implies that the term $\frac{\partial\psi}{\partial r}$ which appears in equation (7) actually belongs to $H^{3/2}(S^1,\mathbb{R}^3)$. (The term $\frac{\partial\psi}{\partial r}$ in equation (7) is in reality $\frac{\partial\psi}{\partial r}$ evaluated at $r = 1$ but hopefully this abuse in notation should cause no confusion.)

Continuing we now see that $\frac{\partial\hat{\Omega}}{\partial r}(u)[X^\alpha(u)] \in H^{3/2}(S^1,\mathbb{R}^3)$. Looking at equation (7) again this tells us that $\frac{\partial X^\alpha}{\partial r}(u) \in H^1(S^1,\mathbb{R}^3)$ or that $X^\alpha(u) \in H^2(S^1,\mathbb{R}^3)$, which concludes the proof of Theorem 5.

Theorem 5 tells us that for each u, $X^\alpha(u)$ is a vector field over u which is in $H^2(S^1,\mathbb{R}^3)$. If we can show that $u \longmapsto X^\alpha(u)$ is C^{r-1} then X^α is a C^{r-1} fector field on the Hilbert manifold $\mathcal{N}_\alpha$.

THEOREM 6. $u \longmapsto X(u)$ is a C^{r-1} map from $\mathcal{N}_\alpha$ to $H^2(S^1,\mathbb{R}^3)$.

Proof. Recall that by a vector field over $u \in H^2(S^1,\Gamma)$ we mean a map $\gamma\colon S^1 \to \mathbb{R}^3$ with $\gamma(\theta) \in T_{u(\theta)}\Gamma$ for almost all θ. If γ is continuous then $\gamma(\theta) \in T_{u(\theta)}\Gamma$ for all θ.

LEMMA. The correspondence $\gamma \to L_u(\gamma)$, $L_u(\gamma) = \Omega(u)\frac{\partial\gamma}{\partial r}$ defines a map from $H^2(S^1,\mathbb{R}^3)$ to $H^1(S^1,\mathbb{R}^3)$ which restricts to an isomorphism between the H^2 and H^1 vector fields over u. The H^2 vector fields over u; i.e., $(T_u\mathcal{N}_\alpha)^1 = \{h \in H^1(S^1,\mathbb{R}^3) \mid h(\theta) \in T_{u(\theta)}\Gamma\}$. Thus $L_u\colon T_u\mathcal{N}_\alpha \to (T_u\mathcal{N}_\alpha)^1$ is an isomorphism.

Proof (of lemma; a sketch).

First, observe that if $\gamma \in T_u\mathcal{N}_\alpha$ with $L_u(\gamma) = 0$ then

$$\int_{S^1} \left\langle L_u(\gamma), \gamma \right\rangle d\theta = \int_{S^1} \left\langle \frac{\partial\gamma}{\partial r}, \gamma \right\rangle d\theta = |||\gamma|||^2 = 0$$

which implies (by the lemma to Theorem 2) that $\gamma = 0$. Thus L_u is an injective map on $T_u\mathcal{N}_\alpha$. To show that L_u is onto let $\phi \in (T_u\mathcal{N}_\alpha)^1$ and define the functional $\Phi_u\colon H_u^{1/2} \to \mathbb{R}$ by

$$\Phi_u(h) = \frac{1}{2}\int_{S^1} \left\langle \frac{\partial h}{\partial r} - \frac{\partial\rho}{\partial r}, h - \rho \right\rangle d\theta = |||h - \rho|||^2.$$

where $\frac{\partial\rho}{\partial r} = \phi$

Using methods completely analogous to those used in the proof in the existence of $X^\alpha(u)$ we find that there exists a unique minimum $\gamma \in H_u^{1/2}$ with $\Omega(u)\frac{\partial\gamma}{\partial u} = \phi$.

Employing the same regularity methods as in the case of $X^\alpha(u)$ we find that $\gamma \in H^2$. Thus L_u is a continuous, linear, bijective map from $T_u\mathcal{N}_\alpha$ to $(T_u\mathcal{N}_\alpha)^1$. From the open mapping theorem it follows that L_u has a continuous linear inverse. This proves the lemma.

Continuing with the proof of the theorem we note that (by Lemma 5 of Section II) the map $u \longmapsto L_u$ is a C^{r-1} map from $\mathcal{N}_\alpha$ to $(H^2(S^1,\mathbb{R}^3), H^1(S^1,\mathbb{R}^3))$ and the map $u \longmapsto \Omega(u)\frac{\partial u}{\partial r}$ is also a C^{r-1} map from $\mathcal{N}_\alpha$ to $H^1(S^1,\mathbb{R}^3)$.

We have seen that L_u restricts to a linear isomorphism of $T_u\mathcal{N}_\alpha$ and $(T_u\mathcal{N}_\alpha)^1$. Letting L_u^{-1} denote the inverse of this map and noting that for fixed u, $L_u \to L_u^{-1}$ is C^∞ we can conclude that $u \longmapsto X^\alpha(u) = L_u^{-1}[\Omega(u)\frac{\partial u}{\partial r}]$ is of class C^{r-1}. Since $\mathcal{N}_\alpha$ is a C^r submanifold of $H^2(S^1,\mathbb{R}^3)$, X is a C^{r-1} vector field on $\mathcal{N}_\alpha$.

We are therefore in the possession of a C^{r-1} vector field X^α which has been obtained in a natural way from the Dirichlet functional. We next show that $X_u^\alpha(E_\alpha) \geq 0$ (X^α is transverse to E_α) and that the zeros

X^α are precisely the critical points of E_α. Thus the zeros of X^α are precisely the simply connected minimal surfaces spanning Γ.

THEOREM 7. $X_u^\alpha(E_\alpha) = DE_\alpha(u)(X^\alpha(u)) \geq 0$ and equals zero iff u is a critical point of E_α.

Proof. The Fréchet derivative of E_α at u is given by

$$DE_\alpha(u)[h] = \int_{S^1} \left\langle \frac{\partial u}{\partial r}, h \right\rangle d\theta = \int_{S^1} \left\langle \Omega(u) \frac{\partial u}{\partial r}, h \right\rangle d\theta .$$

Thus

$$DE_\alpha(u)[X^\alpha(u)] = \int_{S^1} \left\langle \Omega(u) \frac{\partial u}{\partial r}, X^\alpha(u) \right\rangle d\theta$$

$$= \int_{S^1} \left\langle \Omega(u) \frac{\partial X^\alpha}{\partial r}(u), X^\alpha(u) \right\rangle d\theta = \int_{S^1} \left\langle \frac{\partial X^\alpha}{\partial r}(u), X^\alpha(u) \right\rangle d\theta$$

$$= |||X^\alpha(u)|||^2 = 0 \text{ iff } X^\alpha(u) = 0 \text{ iff } \Omega(u) \frac{\partial u}{\partial r} = 0 .$$

We can give another description of the vector field X^α which is in the spirit of [35]. Let $\Gamma = \alpha(S^1)$ and $T, S, R : \Gamma \to \mathbb{R}^3$ be a smooth framing of Γ; i.e., for each $p \in \Gamma$, $T(p)$, $R(p)$, $S(p)$ are an orthonormal basis for $\mathbb{R}^3$ with $T(p) \in T_p\Gamma$. Then the vector field X^α is characterized by the following conditions. For each $u \in \mathcal{N}_\alpha$, $X^\alpha(u)$ is a harmonic map on the disc

(i) $\Delta X^\alpha(u) = 0$

satisfying the mixed Neumann-Dirichlet boundary conditions on S^1

(ii) $\frac{\partial X^\alpha}{\partial r}(u) \cdot T(u) = \frac{\partial u}{\partial r} \cdot T(u)$

(iii) $X^\alpha(u) \cdot S(u) = 0$

(iv) $X^\alpha(u) \cdot R(u) = 0$

where $R(u)$ denotes the composition $R(u(\theta))$, and similarly for S and T.

V. A PROOF THAT THE VECTOR FIELD X^{α}, ASSOCIATED TO E_{α}, IS PALAIS-SMALE

In order to prove that X^{α} is a Palais-Smale vector field on we must first demonstrate the existence of a connection on $\mathcal{N}_{\alpha}$ which will be the candidate connection with respect to which X^{α} will be Palais-Smale. Before doing this let us briefly review the theory of connections on Banach bundles.

For the exposition of connections on such bundles we follow Eliasson [7] and the reader is referred to that source. Let $\pi: \xi \to M$ be a Banach bundle of class C^r with fibre E over a paracompact Banach manifold M (possibly with boundary), modelled on a Banach space F. Let $\lambda : T\xi \to \xi$ denote the tangent bundle to the bundle ξ with canonical projection map λ.

A local trivialization of π is given by a bundle equivalence

$$\begin{array}{ccc} \pi^{-1}(\mathcal{U}) & \xrightarrow{\Phi} & \phi(\mathcal{U}) \times E \\ \downarrow & & \downarrow \\ \mathcal{U} & \xrightarrow{\phi} & \phi(\mathcal{U}) \end{array}$$

where ϕ is a chart for M. $\Phi_p = \Phi|\pi^{-1}(p)$ is a linear isomorphism of $E_p = \pi^{-1}(p)$ onto $\phi(p) \times E \cong E$.

DEFINITION. A connection map K for the bundle ξ is a bundle map $K: T\xi \to \xi$ over the map π. Thus we have the commutative diagram

$$\begin{array}{ccc} T\xi & \xrightarrow{K} & \xi \\ \lambda \downarrow & & \downarrow \pi \\ \xi & \xrightarrow{\pi} & M \end{array}$$

which implies that K maps the fibre $T_z\xi$ linearly into $E_{\pi\lambda(z)}$.

In addition we assume that for any local trivialization $(\mathcal{U},\phi,\Phi)$ of $\pi\colon \xi \to M$ there is a C^{ℓ} map $(\ell \leq r-1)$ $\Gamma_{\phi}\colon \phi(\mathcal{U}) \to \mathcal{L}(E \times F;E)$, the E valued bilinear maps on $E \times F$, which gives the local representative of K, $K_{\phi} = \Phi \circ K \circ T\Phi^{-1}$ by the formula

$$K_{\phi}(x,\rho,y,\eta) = (x,\eta + \Gamma_{\phi}(x)(y,\rho)) .$$

In this case we say that K is of class C^{ℓ}. The map Γ_{ϕ} is the local connector for K with respect to the chart ϕ. In the case M is C^{r+1}, $\xi = TM$ the local connector Γ_{ϕ} corresponds to the classical Christoffel symbols and in finite dimensions we have $[\Gamma_{\phi}(x)(y,z)]^{i} = \Sigma\Gamma^{i}_{jk}(x)y^{j}z^{k}$.

If $\xi = TM$ then we say that K is symmetric if locally there is a chart ϕ so that $\Gamma_{\phi}(x)\colon E \times E \to E$ is a symmetric bilinear map. It is easy to check that this property is independent of the choice of coordinate chart ϕ. It follows easily that if M admits a partition of unity of class C^{r-1} then there exists a C^{r-1} connection map for $\pi\colon \xi \to M$.

A connection for a manifold M is defined to be a connection for its tangent bundle TM. The important thing about bundle connections from our point of view is that they occur naturally in the context of manifolds of maps. This was proved in great generality by Eliasson [7] (Theorem 5.4) and by the author at a later date. It is this fact that motivated the present theory.

"Roughly speaking" if X and Y are two finite dimensional manifolds of class C^{s} with Y admitting a connection K of class C^{s-2} and $\mathcal{G}$ is a manifold of maps functor (e.g., L^{p}_{k}, $k > (\dim X)/p$, C^{k}, $C^{k,\alpha}$) then K induces a C^{s-4} connection $\mathcal{G}(K)$ on the manifold of maps $\mathcal{G}(X,Y)$. If K is symmetric, then $\mathcal{G}(K)$ is also symmetric.

Again, let M be a differentiable Banach manifold with tangent bundle $\tau\colon TM \to M$. In addition, let $\pi\colon \xi \to M$ be a Banach vector bundle with a connection map K on $T\xi$. If $\mathcal{S}\colon M \to \xi$ is a smooth section of ξ we define the covariant derivative of $\mathcal{S}$ to be the bundle map $\nabla\mathcal{S}\colon TM \to \xi$ defined by $\nabla\mathcal{S}(p) = K \circ D\mathcal{S}(p)$. $D\mathcal{S}(p)$, the Fréchet derivative of $\mathcal{S}$ at p, takes the fibre T_pM linearly into $T_{\mathcal{S}(p)}\xi$ and K maps $T_{\mathcal{S}(p)}\xi$ into E_p. Therefore $\nabla\mathcal{S}(p)$ is a linear map from T_pM to E_p.

DEFINITION. A C^k, $k \leq r$, section $\mathcal{S}$ of a C^r tangent bundle $\pi\colon TM \to M$ is said to be Palais-Smale on a set $A \subset M$ with respect to a connection K on TM if for each $p \in A$, the covariant derivative $\nabla\mathcal{S}(p)\colon T_pM \to T_pM$ is an element of $G\mathcal{L}_C(T_pM)$. $\mathcal{S}$ is Palais-Smale with respect to K if it is Palais-Smale on all of M.

If p is a zero of a vector field $\mathcal{S}$, $\mathcal{S}(p) = 0$, then the covariant derivative of $\mathcal{S}$ at p, $\nabla\mathcal{S}(p)$, does not depend on the connection K. To see this, note that in a local coordinate chart ϕ with local representative $\mathcal{S}^\phi$ of $\mathcal{S}$

$$\nabla\mathcal{S}^\phi(p)[h] = D\mathcal{S}^\phi(p)[h] + \Gamma_\phi(p)(h, \mathcal{S}^\phi(p)).$$

Consequently, if $\mathcal{S}(p) = 0$

$$\nabla\mathcal{S}(p)[h] = D\mathcal{S}(p)[h].$$

Therefore the notion of being Palais-Smale on the set of zeros of a vector field does not depend on the choice of connection. However, for not identically zero vector fields the global notion of being Palais-Smale depends strongly on the choice of connection. A non-identically zero vector field could be Palais-Smale with respect to a connection K_1 and yet not Palais-Smale with respect to another connection K_2. One can define an equivalence relation on the space of connections in such a way that the property of being Palais-Smale is independent of the choice of representative connection in any equivalence class. For a detailed discussion see [35].

The manifold $\mathcal{N}_\alpha$ is a submanifold of the linear space of maps $H^2(S^1,\mathbb{R}^3)$. Therefore $T\mathcal{N}_\alpha$ can be thought of as a subbundle of the product bundle $\mathcal{N}_\alpha \times H^2(S^1,\mathbb{R}^3)$ and $T^2\mathcal{N}_\alpha$ as a subbundle of the product bundle

$$T\mathcal{N}_\alpha \times [H^2(S^1,\mathbb{R}^3) \times H^2(S^1,\mathbb{R}^3)] = T\mathcal{N}_\alpha \times [H^2 \times H^2] \quad . \tag{1}$$

Let $p = (u,h) \in T\mathcal{N}_\alpha \subset \mathcal{N}_\alpha \times H^2(S^1,\mathbb{R}^3)$ and $(\tilde{\rho},\tilde{\eta}) \in H^2 \times H^2$ be an element of $T_p(T\mathcal{N}_\alpha)$ in representation (1). For each p we must define a connection $K\colon T_p(T\mathcal{N}_\alpha) \to T_u\mathcal{N}_\alpha$. Let $\Omega\colon \Gamma \to \mathcal{OP}(\mathbb{R}^3)$ be the map introduced in Section III. Define

$$K(u,h,\tilde{\rho},\tilde{\eta}) = (u,\Omega(u)\tilde{\eta}) \quad . \tag{2}$$

To see that K is a connection we must check that K has the proper local form. Let $\phi\colon \mathcal{U} \to V \subset E \subset H^2(S^1,\mathbb{R}^3)$ be a local chart for $\mathcal{N}_\alpha$ about u where E represents the tangent space to $\mathcal{N}_\alpha$ at u. Set $\varphi = \phi^{-1}\colon V \to \mathcal{U}$, $x = \phi(u)$, $\rho = D\phi(u)[h]$, and let $(y,\eta) \in E \times E$. Then $(u,D\varphi(x)[\rho]) \in T_u\mathcal{N}_\alpha$ and

$$(u,D\varphi(x)[\rho],D\varphi(x)[y],D\varphi(x)[\eta] + D^2\varphi(x)[y,\rho])$$

is an element of $T^2\mathcal{N}_\alpha$ in the representation given by (1). Therefore, in local coordinates $K_\phi(x,y,\rho,\eta) = \Omega(u)D\varphi(x)[\eta] + \Omega(u)D^2\varphi(x)[y,\rho] = \eta + \Omega(u)D^2\varphi(x)[y,\rho]$. Thus the local connector $\Gamma_\phi(x)$ (or Christoffel symbol) is given by the formula

$$\Gamma_\phi(x)(y,\rho) = \Omega(u)D^2\varphi(x)[y,\rho] \quad .$$

As mentioned in Section III the map $\Omega: \Gamma \to \mathcal{O}\mathcal{P}(\mathbb{R}^3)$ is C^{r+1} smooth. It follows that $u \longmapsto \Omega(u)$ is C^{r-1} and since $\mathcal{N}_\alpha$ is a C^r manifold it further follows that K is a C^{r-2} connection.

If we give Γ the Riemannian structure it inherits from $\mathbb{R}^3$, then the projection Ω determines (and is determined by) a connection k on the bundle $T\Gamma$; in fact, k is the unique symmetric affine connection associated to this induced Riemannian structure on $T\Gamma$. We will think of K as the connection induced in $T\mathcal{N}_\alpha$ by the connection k on $T\Gamma$. This induced connection is the crucial link between the geometry of Γ as it is embedded in $\mathbb{R}^3$ and the geometry of the manifold of maps $\mathcal{N}_\alpha$. For example the vector field X^α of the last section is determined by K. Let us paraphrase what we have done so far as a

THEOREM 1. Let $\Gamma = \alpha(S^1)$, where $\alpha \in H^{r+2}(S^1,\mathbb{R}^3)$ is an embedding. Let k be the connection induced on $T\Gamma$ by the Riemannian structure $T\Gamma$ inherits from $\mathbb{R}^3$. Then k induces a C^{r-2} symmetric connection K on the C^{r-1} bundle $T\mathcal{N}_\alpha$.

Suppose X^α is the C^{r-1} vector field on the C^r manifold $\mathcal{N}_\alpha$ associated to the Dirichlet functional $\acute{E}_\alpha$. Given the connection K on $\mathcal{N}_\alpha$ we can take the covariant derivative of X^α as follows:

$$X^\alpha: \mathcal{N}_\alpha \to T\mathcal{N}_\alpha \subset \mathcal{N}_\alpha \times H^2(S^1,\mathbb{R}^3) .$$

By taking principal parts we can consider X^α as a map from $\mathcal{N}_\alpha$ to $H^2(S^1,\mathbb{R}^3)$. The Fréchet derivative $X^\alpha_*(u)$ of X at u is a linear map from $T_u\mathcal{N}_\alpha$ to $H^2(S^1,\mathbb{R}^3)$. <u>The covariant derivative of</u> X^α <u>with respect to</u> K <u>is then given by the formula</u>

(3) $$\nabla X^\alpha(u)[h] = \Omega(u)X^\alpha_*(u)[h]$$

for $h \in T_u\mathcal{N}_\alpha$.

Since $\Omega(u)\colon\ H^2(S^1,\mathbb{R}^3) \to T_u\mathcal{N}_\alpha$ is linear we immediately see that $\nabla X^\alpha(u)$ is a linear map of $T_u\mathcal{N}_\alpha$ to itself. We are now prepared to prove the major result of this section.

THEOREM 2. The vector field X^α on $\mathcal{N}_\alpha$ is Palais-Smale with respect to the connection $K\colon\ T^2\mathcal{N}_\alpha \to T\mathcal{N}_\alpha$ given by Theorem 1 .

Proof. By definition

$$\Omega(u)\ \frac{\partial X^\alpha}{\partial r}\ (u) = \Omega(u)\ \frac{\partial u}{\partial r}\ .$$

Thus, if $h \in T_u\mathcal{N}_\alpha$

$$\Omega_*(u)[h]\left[\frac{\partial X^\alpha}{\partial r}\ (u)\right] + \Omega(u)\ \frac{\partial X_*^\alpha}{\partial r}\ (u)[h] = \Omega_*(u)[h]\left[\frac{\partial u}{\partial r}\right] + \Omega(u)\ \frac{\partial h}{\partial r}$$

where for fixed $\beta \in H^2$ $\Omega_*(u)[\cdot][\beta]$ represents the Fréchet derivative of the map $u \to \Omega(u)\beta$ of $\mathcal{N}_\alpha$ to $H^2(S^1,\mathbb{R}^3)$, and where differentiation with respect to r , say $\frac{\partial h}{\partial r}$ denotes the normal derivative of the harmonic extension of h evaluated at $r = 1$. Moreover

$$\Omega_*(u)[h](\theta) = D\Omega(u(\theta))[h(\theta)] \tag{4}$$

where $D\Omega(p)\colon\ T_p\Gamma \to \mathcal{L}(\mathbb{R}^3,\mathbb{R}^3)$ is the finite dimensional Fréchet derivative of $\Omega\colon\ \Gamma \to \mathcal{O}\mathcal{P}(\mathbb{R}^3)$. Continuing we have

$$\Omega(u)\left[\frac{\partial}{\partial r}\ X_*^\alpha(u)[h] - \frac{\partial h}{\partial r}\right] = \Omega_*(u)[h]\left[\frac{\partial u}{\partial r}\right] - \Omega_*(u)[h]\left[\frac{\partial X^\alpha}{\partial r}\ (u)\right]\ . \tag{5}$$

LEMMA 1. The map $\mathcal{R}_u\colon\ T_u\mathcal{N}_\alpha \to (T_u\mathcal{N}_\alpha)^1$ given by

$$h \xmapsto{\ \mathcal{R}_u\ } \Omega_*(u)[h]\left[\frac{\partial u}{\partial r}\right] - \Omega_*(u)[h]\left[\frac{\partial X}{\partial r}(u)\right]$$

is completely continuous (or compact).

Proof (of lemma). We know that $u \mapsto \Omega(u)$ is C^{r-1}, $r \geq 5$. Also $u \to D\Omega(u(\theta))$ is C^{r-2}. Theorems 3 and 5 of Section II imply that $\mathcal{R}_u$ maps $T_u\mathcal{N}_\alpha$ continuously to $H^1(S^1, \mathbb{R}^3)$. We must show that if $\{h_n\}$ is a bounded sequence in $T_u\mathcal{N}_\alpha$ (and hence bounded in $H^2(S^1,\mathbb{R}^3)$) that $\{h_n\}$ has a subsequence $\{h_{n_j}\}$ with $\{\mathcal{R}(h_{n_j})\}$ converging in $H^1(S^1,\mathbb{R}^3)$.

In the last section we introduced the notation $(T_u\mathcal{N}_\alpha)^1$ to represent the H^1 vector fields over u. Clearly $T_u\mathcal{N}_\alpha \subset (T_u\mathcal{N}_\alpha)^1$. One readily check that $\mathcal{R}_u$ extends to a continuous linear map of $(T_u\mathcal{N}_\alpha)^1$ to $H^1(S^1,\mathbb{R}^3)$. By the Sobelev embedding theorem if $\{h_n\}$ is a bounded sequence in $H^2(S^1,\mathbb{R}^3)$ then $\{h_n\}$ has a subsequence $\{h_{n_j}\}$ which converges in H^1 to an element $h \in H^1$. Since $\mathcal{R}_u$ extends to a continuous map on $(T_u\mathcal{N}_\alpha)^1$ $\mathcal{R}_u(h_{n_j})$ converges to $\mathcal{R}_u(h)$. This completes the proof of the lemma.

Let $\gamma \in T_u\mathcal{N}_\alpha$. By the lemma in the proof of Theorem 6 Section IV the map $L_u\colon T_u\mathcal{N}_\alpha \to (T_u\mathcal{N}_\alpha)^1$ given by $L_u\gamma = \Omega(u)\dfrac{\partial\gamma}{\partial r}$ is an isomorphism. $X_*^\alpha(u)[h] - h$ need not be a vector field over u although if u is a zero of X it will be. The projection $\Omega(u)\{X_*(u)[h] - h\}$ is an H^2 vector field over u and hence an element of $T_u\mathcal{N}_\alpha$.

LEMMA 2.

$$(6)\qquad L_u\Omega(u)\{X_*(u)[h] - h\} = \Omega(u)\frac{\partial}{\partial r}\{X_*(u)[h] - h\} + A_u h$$

where $A_u : T_u\mathcal{N}_\alpha \to (T_u\mathcal{N}_\alpha)^1$ is a compact linear map.

Proof (of Lemma 2). Let $\gamma \in H^2(S^1,\mathbb{R}^3)$. Then set $\dfrac{\partial}{\partial r}\Omega(u)\gamma = \dfrac{\partial}{\partial r}\widehat{\Omega(u)\gamma}\Big|_{r=1}$ where $\widehat{\Omega(u)\gamma}$ is the harmonic extension of $\Omega(u)\gamma$ to the unit disc $\mathcal{D}$ {cf. Section IV}.

If in Theorem 4 of Section IV we replace $X^\alpha(u)$ by $\gamma \in H^2(S^1,\mathbb{R}^3)$ the same proofs show that

$$\hat{\Omega}(u)\hat{\gamma} = \widehat{\Omega(u)\gamma} + \psi(\gamma) \tag{7}$$

where $\psi(\gamma) \in H^3(\mathcal{D},\mathbb{R}^3)$ satisfies $\psi(\gamma)|S^1 = 0$ and $\Delta\psi(\gamma) = 2\nabla\hat{\Omega}(u)\cdot\nabla\hat{\gamma}$. Clearly the dependence of $\psi(\gamma)$ on γ is linear in γ. Differentiating (7) with respect to r and setting $r = 1$ we obtain

$$\frac{\partial}{\partial r}\{\hat{\Omega}(u)\hat{\gamma}\} = \frac{\partial}{\partial r}\{\Omega(u)\gamma\} + \frac{\partial}{\partial r}\psi(\gamma) \quad . \tag{8}$$

Moreover on the disc

$$\frac{\partial}{\partial r}\{\hat{\Omega}(u)\hat{\gamma}\} = \frac{\partial\hat{\Omega}}{\partial r}(u)[\hat{\gamma}] + \hat{\Omega}(u)\left[\frac{\partial\hat{\gamma}}{\partial r}\right]$$

and so (8) can be rewritten (continuing our convention of dropping the ^ when differentiating w.r.t. r, r = 1) as

$$\frac{\partial}{\partial r}\{\Omega(u)\gamma\} = \Omega(u)\left[\frac{\partial\gamma}{\partial r}\right] + \frac{\partial\Omega}{\partial r}(u)[\gamma] - \frac{\partial}{\partial r}\psi(\gamma) \tag{9}$$

where $\frac{\partial}{\partial r}\psi(\gamma)|S^1 \in H^{3/2}(S^1,\mathbb{R}^3)$. Setting $\gamma = \{X_*(u)[h] - h\}$ in equation (9) we obtain

$$\frac{\partial}{\partial r}[\Omega(u)\; X_*(u)[h]-h\}] = \Omega(u)\frac{\partial}{\partial r}\{X_*(u)[h]-h\} + \frac{\partial\Omega}{\partial r}(u)[\gamma] - \frac{\partial}{\partial r}\psi(\gamma) \tag{10}$$

where $\frac{\partial\Omega}{\partial r}(u)[\gamma] \in H^1(S^1,\mathbb{R}^3)$.

Multiplying both sides of (10) by $\Omega(u)$ we get equation (6) where

$$A_u(h) = \Omega(u)\frac{\partial\Omega}{\partial r}(u)[\gamma] - \Omega(u)\frac{\partial}{\partial r}\psi(\gamma) \; ,$$

$\gamma = \{X_*(u)[h] - h\}$. The map $h \longmapsto \Omega(u)\frac{\partial\Omega}{\partial r}(u)[\gamma]$ extends to a continuous linear map of $(T_u\mathcal{N}_\alpha)^1$ to itself and so as in Lemma 1 it is compact linear. The map $h \mapsto \Omega(u) \frac{\partial}{\partial r} \psi(\gamma)$ is a continuous linear map of $T_u\mathcal{N}_\alpha$ into $(T_u\mathcal{N}_\alpha)^{3/2}$, the $H^{3/2}$ vector fields over u. Since the inclusion of $H^{3/2}$ into H^1 is compact linear it follows that the linear map $h \longmapsto \Omega(u) \frac{\partial}{\partial r} \psi(\gamma)$ as a map into $(T_u\mathcal{N}_\alpha)^1$ is compact. Thus A_u is compact which concludes the proof of the lemma.

We are now in a position to complete the proof of Theorem 2 . Repeating equation (5) we have

$$\Omega(u)\left[\frac{\partial}{\partial r} X_*(u)[h] - \frac{\partial h}{\partial r}\right] = \mathcal{R}_u(h) \quad .$$

Using (6) we get

$$L_u\Omega(u)\{X_*^\alpha(u)[h] - h\} = \mathcal{R}_u(h) + A_u(h)$$

where both A_u and $\mathcal{R}_u$ are compact linear maps. Consequently

$$\begin{aligned}\Omega(u)\{X_*^\alpha(u)[h] - h\} &= L_u^{-1}\{\mathcal{R}_u(h) + A_u(h)\}\\ &= S_u(h)\end{aligned}$$

or using (3)

$$\nabla X^\alpha(u)[h] = h + S_u(h)$$

where $S_u\colon T_u\mathcal{N}_\alpha \hookleftarrow$ is a compact linear map. Thus $h \longmapsto h + S_u(h)$ is an element of $\mathcal{L}_C(T_u\mathcal{N}_\alpha)$. This shows that X is Palais-Smale with respect to the connection K given by Theorem 1.

In Section IX we shall derive an explicit expression for $\nabla X^\alpha(a)$ at a zero u in terms of geometrical data of Γ . We end this section with an interesting corollary to Theorem 2. The proof can be found in [35] .

COROLLARY. Let $E_\alpha : \mathcal{N}_\alpha \to R$ be the Dirichlet energy functional. Then the critical points of E_α (which are the simply connected minimal surfaces spanning Γ) are all of finite Morse index. That is, for any critical point u, the maximal subspace of $T_u \mathcal{N}_\alpha$ on which the Hessian $D^2E_\alpha(u): T_u \mathcal{N}_\alpha \times T_u \mathcal{N}_\alpha \to \mathcal{R}$ is negative definite is of finite dimension.

VI. THE WEAK RIEMANNIAN STRUCTURE ON $\mathcal{N}_\alpha$.

In this section we (a) introduce the weak Riemannian structure on $\mathcal{N}_\alpha$, (b) show that for each $u \in \mathcal{N}_\alpha$, $H^2(S^1,\mathbb{R}^3)$ has a splitting $H^2(S^1,\mathbb{R}^3) = T_u\mathcal{N}_\alpha \oplus T_u\mathcal{N}_\alpha^\perp$ where $T_u\mathcal{N}_\alpha^\perp$, (c) interpret the vector field X as the gradient of E_α with respect to this weak structure, and finally (d) give a new formula for X in terms of a weak orthogonal projection.

DEFINITION. Let $u \in \mathcal{N}_\alpha$ and $h,k \in T_u\mathcal{N}_\alpha$. Define $\langle\langle h,k\rangle\rangle_u$, the weak inner product of h and k over u by the formula

$$\langle\langle h,k\rangle\rangle_u = \int_{S^1}\left\langle\frac{\partial h}{\partial r}, k\right\rangle d\theta .$$

The weak norm $|||\cdot|||^2 = \langle\langle\cdot,\cdot\rangle\rangle$ was introducted in Section IV, where we saw that $\langle\langle h,h\rangle\rangle_u = 0$ iff $h = 0$.

Let $\Delta^*(T\mathcal{N}_\alpha \times T\mathcal{N}_\alpha)$ denote the pull back of the product bundle $T\mathcal{N}_\alpha \times T\mathcal{N}_\alpha$ over $\mathcal{N}_\alpha \times \mathcal{N}_\alpha$ via the diagonal map $\Delta : \mathcal{N}_\alpha \to \mathcal{N}_\alpha \times \mathcal{N}_\alpha$. $\Delta^*(T\mathcal{N}_\alpha \times T\mathcal{N}_\alpha)$ is a subbundle of the product bundle $\mathcal{N}_\alpha \times [H^2(S^1,\mathbb{R}^3) \times \times H^2(S^1,\mathbb{R}^3)]$. $\langle\langle,\rangle\rangle : \Delta^*(T\mathcal{N}_\alpha \times T\mathcal{N}_\alpha) \to R$ is the restriction of a C^r fibrewise bilinear form on this product bundle, and so $\langle\langle,\rangle\rangle$ is a C^r map which is bilinear on each fibre.

DEFINITION. Let $F \subset H^2(S^1,\mathbb{R}^3)$ be a closed subspace. By the weak orthogonal complement $F^\perp$ of F we mean

$$\{h \in H^2(S^1,\mathbb{R}^3) | \langle\langle h,k\rangle\rangle = 0 \quad k \in F\} .$$

In general it will <u>not</u> be true that $H^2(S^1,\mathbb{R}^3) = F \oplus F^\perp$. In fact F could be a closed proper subspace of $H^2(S^1,\mathbb{R}^3)$ with $F^\perp = 0$. The next result shows that because of the naturality of the Plateau problem $T_u\mathcal{N}_\alpha \subset H^2(S^1,\mathbb{R}^3)$ does have a closed complement in $H^2(S^1,\mathbb{R}^3)$.

THEOREM 1. Let $u \in \mathcal{N}_\alpha$. Then

$$H^2(S^1, \mathbb{R}^3) = T_u\mathcal{N}_\alpha \oplus T_u\mathcal{N}_\alpha^\perp .$$

Proof. For $h \in H^2(S^1, \mathbb{R}^3)$ let $L_u\gamma = \Omega(u)\dfrac{\partial h}{\partial r}$. In Section IV we showed that L_u restricts to an isomorphism from $T_u\mathcal{N}_\alpha$ to $(T_u\mathcal{N}_\alpha)^1$. Thus there is a unique $h_T \in T_u\mathcal{N}_\alpha$ with $L_u h_T = \Omega(u)\dfrac{\partial h}{\partial r}$. Clearly

$$h = h_T + (h - h_T). \tag{1}$$

We claim that $h - h_T$ is weak orthogonal to h_T, for

$$\langle\langle h_T, h - h_T\rangle\rangle = \int_{S^1}\left\langle h_T, \frac{\partial}{\partial r}(h - h_T)\right\rangle d\theta = \int_{S^1}\left\langle h_T, \Omega(u)\frac{\partial}{\partial r}h - \Omega(u)\frac{\partial h_T}{\partial r}\right\rangle d\theta = 0.$$

Therefore (1) gives us the desired decompostion of $H^2(S^1, \mathbb{R}^3)$.

Let us now interpret the vector field X^α in terms of this weak Riemannian structure. Recall that the Dirichlet functional $E_\alpha : \mathcal{N}_\alpha \to R$ is given by

$$E_\alpha(u) = \frac{1}{2}\int_{S^1}\left\langle\frac{\partial u}{\partial r}, u\right\rangle d\theta$$

with

$$DE_\alpha(u)[h] = \int_{S^1}\left\langle\frac{\partial u}{\partial r}, h\right\rangle d\theta = \int_{S^1}\left\langle\Omega(u)\frac{\partial u}{\partial r}, h\right\rangle d\theta = \langle\langle X^\alpha(u), h\rangle\rangle_u .$$

Thus $DE_\alpha(u)[h] = \langle\langle X^\alpha(u), h\rangle\rangle_u$ which means that X^α <u>is the gradient of E_α with respect to</u> $\langle\langle\, ,\rangle\rangle$.

It follows from Theorem 1 that for each $u \in \mathcal{N}_\alpha$ there is a weak orthogonal projection $\Pi_u : H^2(S^1, \mathbb{R}^3) \to T_u\mathcal{N}_\alpha$ which sends h to h_T. We conclude this section with

THEOREM 2. The weak gradient vector field X^α of the Dirichlet functional E is given by the equation $X^\alpha(u) = \Pi_u(u)$.

Proof. From the proof of Theorem 1, $u_T = \Pi_u(u)$ is the unique $u_T \in T_u\mathcal{N}_\alpha$ satisfying $\Omega(u)\dfrac{\partial u_T}{\partial r} = \Omega(u)\dfrac{\partial u}{\partial r}$. But $X^\alpha(u)$ also satisfies this equation and so $u_T = X^\alpha(u)$.

§VII. THE EQUIVARIANCE OF X^α UNDER THE ACTION OF THE CONFORMAL GROUP

In this section we introduce the group G of conformal transformations of the disc onto itself. G acts on $\mathcal{N}_\alpha$ in a natural way. We shall show that the weak-Riemannian structure, the Dirichlet functional E_α, and the gradient vector field X^α are all equivariant under the action of G. Moreover G is the maximal transformation group which leaves either E_α or $\langle\langle , \rangle\rangle$ invariant. For a dense set of $u \in \mathcal{N}_\alpha$ the orbit of u under G is an immersion into $\mathcal{N}_\alpha$. If u represents a classical minimal surface; i.e. $u \in \overline{\mathcal{M}}_\alpha$ and a critical point of E_α then the orbit of u is an injective immersion of G into $\mathcal{N}_\alpha$. The group G does not act smoothly on $\mathcal{N}_\alpha$.

Let $\mathcal{U}$ and V be open subsets of the complex plane $\mathbb{C}$. $f : \mathcal{U} \to V$ is <u>conformal</u> if it is complex analytic and has a complex analytic inverse. Let G be the group of conformal maps of the unit disc onto itself. It is well known that every $g \in G$ is of the form

$$g(z) = c\,\frac{z-a}{1-\bar{a}z}$$

where $|c| = 1$ and $|a| < 1$. G is a three dimensional Lie group which is topologically $S^1 \times \mathcal{D}^0$, $\mathcal{D}^0$ the interior of the unit disc. Therefore G is not compact. G acts on $H^2(S^1, \mathbb{R}^3)$ and $\mathcal{N}_\alpha$ in the following manner. Let $u \in H^2(S^1, \mathbb{R}^3)$ and $\hat{u}$ its harmonic extension. For $g \in G$ define $g_\#(\hat{u}) \in H^2(\mathcal{D}, \mathbb{R}^3)$ by

(1) $$g_\#(\hat{u})(z) = \hat{u}(g(z))$$

and

(2) $$g_\#(u)(\Theta) = \hat{u}(g(1, \Theta)) = u(g(\Theta)) .$$

For fixed g, $g_\#$ is a linear isomorphism of H^2 to itself which fixes $\mathcal{N}_\alpha$. Thus $g_\#$ induces a diffeomorphism of $\mathcal{N}_\alpha$ to itself. Again denote this diffeomorphism by $g_\#$. The correspondence $G \to \text{Diff}(\mathcal{N}_\alpha)$ given by $g \to g_\#$ defines the action of G on $\mathcal{N}_\alpha$.

We have already noted that the map $u \longmapsto g_\#(u)$ is a linear isomorphism on $H^2(S^1, \mathbb{R}^3)$. In the case of H^2 let us call this isomorphism A_g, $A_g(u) = g_\#(u)$, since we would like to make a notational distinction between the induced G action on $\mathcal{N}_\alpha (g \longmapsto g_\#)$ and the H^2 representation of G given by $g \longmapsto A_g$.

The action of G on $\mathcal{N}_\alpha$ is <u>not smooth</u> and this, as in the case of closed geodesics on Riemannian manifolds creates a manifold of mathematical difficulties. To see that the action is not smooth first note that $S^1 \simeq \{g \in G \mid g(z) = cz,\ |c| = 1\}$ embeds as a closed subgroup of G and we identify this closed subgroup with S^1. The tangent space to the identity of S^1 is the set $\{\alpha i e^{i\theta} \mid \alpha \in R,\ 0 \le \theta \le 2\pi\}$. This tangent space can be identified with the Lie algebra R of real numbers via the correspondence $\alpha i e^{i\theta} \to \alpha$.

Let $s \longmapsto \phi(s)$ be a C^∞ function on R with $\phi(0) = 0$, and $\phi'(0) = \alpha$. Define a path $s \longmapsto \sigma(s)$ in S^1 by $\sigma(s) = \ell^{i\phi(s)}$. Let $u \in H^2(S^1, \mathbb{R}^3)$. Rather than considering the orbit of u under the full conformal group G, let us look at the orbit of u under S^1. If the action of G were smooth the action of S^1 would be smooth. This in turn would imply that the map $s \longmapsto u(\sigma(s))(\theta) = u(\theta + \phi(s))$ is a smooth map of R into $H^2(S^1, R^3)$. But

$$\frac{d}{ds} u(\sigma(s))\big|_{s=0}(\theta) = u_\theta = \frac{\partial u}{\partial \theta} \in H^1(S^1, \mathbb{R}^3) .$$

Therefore the derivative of this path lies in $H^1(S^1, \mathbb{R}^3)$ and not in $H^2(S^1, \mathbb{R}^3)$ which contradicts the assumption that $s \longmapsto u(\sigma(s))$ is smooth into $H^2(S^1, \mathbb{R}^3)$. Thus the action of G cannot be smooth.

Observe, on the other hand, that if $u \in H^3(S^1, \mathbb{R}^3)$ then $s \mapsto u(\sigma(s))$ is a smooth map of R into $H^2(S^1, \mathbb{R}^3)$. Let $\mathcal{P}_\alpha \subset \mathcal{N}_\alpha$ be the intersection of $\mathcal{N}_\alpha$ with $H^3(S^1, \mathbb{R}^3)$. <u>If</u> $u \in \mathcal{P}_\alpha$ <u>then</u> $g \mapsto g_\#(u)$ <u>is a smooth map into</u> $\mathcal{N}_\alpha$. This follows from Theorem 5, §II. Before showing that this map is an immersion we shall first establish a result which will be of great importance later on, namely, the characterization of the tangent space $T_{id}G$ to G at the identity of G.

THEOREM 1. $T_{id}G$ can be identified with the three dimensional space over $\mathbb{R}$ of analytic maps on the unit disc spanned by $h_1(z) = iz$, $h_2(z) = z^2 - 1$, and $h_3(z) = iz^2 + 1$. Here i denotes $\sqrt{-1}$. These maps restricted to $S^1 = \partial\mathcal{D}$ are tangent to S^1.

<u>Proof</u>. Each element $g \in G$ is of the form $g(z) = c\,\frac{z-a}{1-\bar{a}z}$. Let σ be a C^∞ function R with $\sigma(0) = 0$ and $\sigma'(0) = 1$. Define $c : R \to \mathbb{C}$ by $c(t) = e^{i\sigma(t)}$ and $g_t^1(z) = c(t)z$. Let a_2 and a_3 be C^∞ functions from R to $\mathcal{D}$ with $a_j(0) = 0$ and $a_1'(0) = 1$, $a_2'(0) = -i$. Let

$$g_t^j(z) = \frac{z-a_j(t)}{1-\bar{a}_j(t)z} \quad j = 2, 3 .$$

Then $t \to g_t^j$ are smooth paths in G with $g_0^j = id$ for all j. Therefore $h_j = \frac{d}{dt} g_t^j\big|_{t=0} \in T_{id}G$. A short computation shows that $h_1(z) = iz$, $h_2(z) = z^2 - 1$, and $h_3(z) = iz^2 + 1$. These are linearly independent over $\mathbb{R}$ and since G is a three dimensional Lie group these maps form a basis for $T_{id}G$.

On S^1, $|z| = 1$, $z = e^{i\theta}$ and $h_1(z) = ie^{i\theta}$ which is clearly tangent to S^1 at $e^{i\theta}$; in fact, $ie^{i\theta}$ is a unit tangent vector to S^1 at $e^{i\theta}$. Another simple computation shows that for $z = e^{i\theta}$,

$$h_2(z) = (i\bar{z} - iz)h_1(z) = 2\sin\theta \cdot h_1(z),\ h_3(z) = (z + \bar{z})h_1(z) = 2\cos\theta \cdot h_1(z) .$$

We see that for each $z = e^{i\theta}$, $h_2(z)$ and $h_3(z)$ are real scalar multiples of $h_1(z)$ and thus are tangent to S^1 at $e^{i\theta}$ which ends the proof.

The last part of the above theorem (the tangency to S^1) is a consequence of the fact that G is naturally contained as a subgroup $\hat{G}$ of the C^∞ diffeomorphism group of S^1 since each g fixes S^1.

COROLLARY. The tangent space to $\hat{G}$ at the identity is spanned by the following three linearly independent vector fields along the identity map, $\mathcal{T}(\theta) = ie^{i\theta}$, $\sin\theta \cdot \mathcal{T}(\theta)$, and $\cos\theta \cdot \mathcal{T}(\theta)$, where $\mathcal{T}$ is the oriented unit tangent vector field on S^1. By taking linear combinations it also follows from this that we may take $\mathcal{T}(\theta)$, $e^{i\theta}\mathcal{T}(\theta)$, and $e^{-i\theta}\mathcal{T}(\theta)$ as a basis for $T_{id}\hat{G}$.

THEOREM. Let $u \in \mathcal{P}_\alpha$. Then the orbit $g \to g_\#(u)$ of u under G is an immersed three manifold. If $u \in \bar{\mathcal{M}}_\alpha$ is a critical point of E_α (u is a classical minimal surface) then the orbit of u is a submanifold (the image of an injective immersion) of $\mathcal{N}_\alpha$.

Proof. We have already noted that $g \longmapsto g_\#(u)$ is smooth. To see that this map is an immersion it suffices to check that its derivative at $id \in G$ is non-zero. Consider G as a subgroup $\hat{G}$ of $\mathrm{Diff}(S^1)$ as in the spirit of the last corollary. Then the derivative of $g_\#(u)$ at the identity operating on a vector field $h \in T_{id}\hat{G}$ is given by

$$D[g_\#(u)](id)(h) = Du(\theta)[h(\theta)]$$

where $Du(\theta) : T_{e^{i\theta}}S^1 \to R^3$ denotes the Fréchet derivative of $u : S^1 \to \mathbb{R}^3$ at $e^{i\theta}$. h is a linear combination over $\mathbb{R}$ of the three functions $\mathcal{T}(\theta)$, $\sin\theta \cdot \mathcal{T}(\theta)$ and $\cos\theta \cdot \mathcal{T}(\theta)$. But $u_\theta = \frac{\partial u}{\partial\theta} = Du(\theta)[\mathcal{T}(\theta)]$. Therefore if $h(\theta) = \alpha\mathcal{T}(\theta) + \beta\sin\theta\,\mathcal{T}(\theta) + \rho\cos\theta\,\mathcal{T}(\theta)$, $\alpha, \beta, \rho \in \mathbb{R}$, then

(3) $$D[g_\#(u)](id)(h) = \alpha u_\theta + \beta\sin\theta\, u_\theta + \rho\cos\theta\, u_\theta .$$

Suppose the left-hand side of (3) vanishes. Since u_θ is not identically zero this says that 1, $\sin\theta$, and $\cos\theta$ are linearly depending over some non-empty subset of $[0, 2\pi]$ which is clearly impossible. Consequently $g \longmapsto g_{\#}(u)$ is an immersion.

If $u \in \bar{\mathcal{M}}_\alpha$ is a critical point of E_α then by Theorem 3, §III, u is a homeomorphism of S^1 onto Γ. Hence $g^1_{\#}(u) = g^2_{\#}(u)$ implies that $u(g^1(\theta)) = u(g^2(\theta))$ for all θ which further implies that $g^1 = g^2$, and so $g \to g_{\#}(u)$ is injective. From the regularity results of Hildebrandt and others (cf. §VIII) if $u \in \bar{\mathcal{M}}_\alpha$ is a critical point of E_α then $u \in \mathcal{P}_\alpha$ and by the first part of this theorem $g \to g_{\#}(u)$ is an injective immersion.

As a consequence of the proof of the above theorem we have

COROLLARY. If $u \in \mathcal{P}_\alpha$ the tangent space to the orbit of the group action $g \longmapsto g_{\#}(u)$ is spanned by u_θ, $\sin\theta u_\theta$, and $\cos\theta u_\theta$.

We are now prepared to move onto a proof of the equivariance of the weak Riemannian structure, E_α, and the vector field X^α under the action of G.

Recall that the weak Riemannian structure $\mathcal{N}_\alpha$ inherits from $H^{1/2}(S^1, \mathbb{R}^3)$ is defined by

$$\ll h, k \gg_u = \int_{S^1} \left\langle \frac{\partial h}{\partial r'}, k \right\rangle d\theta$$

for $u \in \mathcal{N}_\alpha$, $h, k \in T_u\mathcal{N}_\alpha$.

The G-action on $\mathcal{N}_\alpha$ induces an action of G on $T\mathcal{N}_\alpha$ as follows. Let $u \in \mathcal{N}_\alpha$ and $h, k \in T_u\mathcal{N}_\alpha$. For <u>fixed</u> g consider the diffeomorphism of $\mathcal{N}_\alpha$ given $u \to g_{\#}(u)$. Let $Dg_{\#}(u)[h]$ denote the Fréchet derivative of this map at a point $u \in \mathcal{N}_\alpha$ evaluated on a tangent vector $h \in T_u\mathcal{N}_\alpha$. The weak structure on $T\mathcal{N}_\alpha$ is invariant under a diffeomorphism ψ of $\mathcal{N}_\alpha$ if for all $u \in \mathcal{N}_\alpha$ $\ll D\psi(u)[h], D\psi(u)[k]\gg = \ll h, k\gg$ for all $h, k \in T_u\mathcal{N}_\alpha$. The weak structure is <u>equivariant</u> under the action of G if for each $g \in G$, $\ll\ \gg$ is invariant under $g_{\#}$.

Let $g \longmapsto A_g$ be the H^2 representation of G introduced at the beginning of this section. It follows easily that the linear isomorphism A_g on $H^2(S^1, \mathbb{R}^3)$ restricts to an isomorphism from $T_u\mathcal{N}_\alpha$ to $T_{g_\#(u)}\mathcal{N}_\alpha$ and moreover that the derivative of $g_\#$ at u, $Dg_\#(u)$ is precisely the linear isomorphism A_g restricted to $T_u\mathcal{N}_\alpha$. Thus the weak structure on $T\mathcal{N}_\alpha$ is equivariant with respect to G if

$$\langle\langle A_g h, A_g k\rangle\rangle = \langle\langle h, k\rangle\rangle \tag{4}$$

for all $h, k \in T_u\mathcal{N}_\alpha$. Explicitly (4) means that for all $g \in G$ and all $h, k \in T_u\mathcal{N}_\alpha$

$$\int_{S^1} \left\langle \frac{\partial}{\partial r}(A_g h), A_g k \right\rangle d\theta = \int_{S^1} \left\langle \frac{\partial h}{\partial r}, k \right\rangle d\theta . \tag{5}$$

The proof of the next result requires only elementary calculus.

THEOREM 3. The weak structure on $T\mathcal{N}_\alpha$ is equivariant under the action of the conformal group.

The Dirichlet functional E_α is invariant under a diffeomorphism if $E_\alpha(\psi(u)) = E_\alpha(u)$ for all $u \in \mathcal{N}_\alpha$. E_α is <u>equivariant</u> with respect to the G action on $\mathcal{N}_\alpha$ if E_α is invariant under $g_\#$ for all $g \in G$.

Another straightforward exercise in the chain rule shows that

THEOREM 4. The Dirichlet functional E_α is equivariant with respect to the action of the conformal group G.

We have already observed that the vector field X^α is the gradient with respect to the weak Riemannian structure. Since $E : \mathcal{N}_\alpha \to R$ and $\langle\langle\ ,\ \rangle\rangle : T\mathcal{N}_\alpha \times T\mathcal{N}_\alpha \to R$ are equivariant with respect to G (or $\hat{G}$) it is reasonable to expect that X^α will be equivariant with respect to $\hat{G}$; that is, if $g \in G$ and $g_\#$ is the induced diffeomorphism of $\mathcal{N}_\alpha$ then

$(Dg_{\#}(u))^{-1}X^{\alpha}(g_{\#}(u)) = X^{\alpha}(u)$ for all $u \in \mathcal{N}_{\alpha}$. This is proven in the next

THEOREM 5. The vector field X^{α} is equivariant with respect to the G action on $\mathcal{N}_{\alpha}$.

Proof. Recall from §VI that for $u \in \mathcal{N}_{\alpha}$ $DE_{\alpha}(u) = \langle\langle X^{\alpha}(u), h\rangle\rangle$ for all $h \in T_n\mathcal{N}_{\alpha}$. Since E_{α} is equivariant the derivative of E_{α} at the point $g_{\#}(u)$ evaluated on the vector $Dg_{\#}(u)[h]$ is given by

$$DE_{\alpha}(g_{\#}(u))[Dg_{\#}(u)[h]] = DE_{\alpha}(u)[h] = \langle\langle X^{\alpha}(u), h\rangle\rangle .$$

But the left-hand side of this equation is equal to $\langle\langle X^{\alpha}(g_{\#}(u)), Dg_{\#}(u)[h]\rangle\rangle$, which by the equivariance of $\langle\langle , \rangle\rangle$ is equal to $\langle\langle [Dg_{\#}(u)]^{-1}X^{\alpha}(g_{\#}(u)), h\rangle\rangle$. Equating terms we see that

$$\langle\langle [Dg_{\#}(u)]^{-1}X^{\alpha}(g_{\#}(u)), h\rangle\rangle = \langle\langle X^{\alpha}(u), h\rangle\rangle$$

for all $h \in T_u\mathcal{N}_{\alpha}$. This implies that $[Dg_{\#}(u)]^{-1}X^{\alpha}(g_{\#}(u)) = X^{\alpha}(u)$ which is precisely what we want to show.

G is a subgroup of the C^1 diffeomorphism group $\mathrm{Diff}^1(\mathcal{D})$ of the disc which preserves orientation. Every subgroup $\mathcal{H} \subset \mathrm{Diff}^1(\mathcal{D})$ induces an action on $\mathcal{N}_{\alpha}$ by composition on the right in exactly the same manner as does G. A natural question is whether there is some larger orientation preserving subgroup $\mathcal{H} \supset G$, or perhaps a completely different subgroup of $\mathrm{Diff}^1(\mathcal{D})$, which leaves the bilinear map $\langle\langle , \rangle\rangle$ on $H^1(\mathcal{D}, \mathbb{R}^3)$ given by $\langle\langle h, k\rangle\rangle = \sum_i \int_{\mathcal{O}} \langle \nabla h^i, \nabla k^i\rangle_{\mathbb{R}^2}$ or the Dirichlet functional E on $H^1(\mathcal{D}, \mathbb{R}^3)$, $E(u) = 1/2\langle\langle u, u\rangle\rangle$, invariant. The next result shows that this cannot happen.

THEOREM 6. Let $\mathcal{H}$ be a maximal orientation preserving subgroup of $\mathrm{Diff}^1(\mathcal{D})$ such that E is equivariant with respect to the induced action on $H^1(\mathcal{D}, \mathbb{R}^3)$. Then $\mathcal{H} = G$.

Proof. Since $\langle\langle h, k\rangle\rangle = \frac{1}{2}\{E(h+k) - E(h-k)\}$ any subgroup of the diffeomorphism group which leaves $\langle\langle , \rangle\rangle$ invariant will leave E invariant and vice versa. Consequently it suffices to consider the case of E only. One can do this directly by assuming that if for a diffeomorphism $\psi \in \mathcal{H}$, $E(u) = E(u \cdot \psi)$ for all $u \in H^1(\mathcal{D}, \mathbb{R}^3)$ then ψ is conformal.

It is easier, however, to note that the equivariance condition implies that the $\mathcal{H}$ action fixes the harmonic maps. Since each harmonic map has a harmonic conjugate we can conclude that $\mathcal{H}$ takes (by composition on the right) analytic maps to analytic maps and therefore must consist of conformal diffeomorphisms of the unit disc, or in other words $\mathcal{H} = G$.

In the Plateau problem we are faced with a more subtle type of invariance than just group actions induced on $H^2(\mathcal{D}, \mathbb{R}^3)$. Since we assume that the maps we consider are harmonic to begin with the Dirichlet integral reduces to

$$E(u) = \frac{1}{2}\int_{S^1}\left\langle \frac{\partial u}{\partial r}, u\right\rangle d\theta \tag{6}$$

and the conformal group G embeds as the subgroup $\hat{G} \subset \mathrm{Diff}^\infty(S^1)$ of the C^∞ diffeomorphisms of S^1 by restriction of the elements of G to the boundary S^1 of $\mathcal{D}$. One can ask if $\hat{G}$ is the maximal subgroup of the diffeomorphism group of S^1 which leaves the functional E and the bilinear pairing $\langle\langle , \rangle\rangle$ on the space $H^2(S^1, \mathbb{R}^3)$ invariant. That the answer is affirmative is not surprising. We have

THEOREM 7. Let $\mathcal{H} \subset \mathrm{Diff}^\infty(S^1)$ be a maximal orientation preserving subgroup which leaves the Dirichlet functional E and the bilinear pairing $\langle\langle , \rangle\rangle$ on $H^2(S^1, \mathbb{R}^3)$ invariant. Then $\mathcal{H} = \hat{G}$.

Proof. It suffices, by earlier remarks, to show that if $\mathcal{H}$ leaves E invariant then $\mathcal{H} = \hat{G}$. $H^2(S^1, R)$ embeds naturally as a closed subspace of

$H^2(S^1, \mathbb{R}^3)$ since $H^2(S^1, \mathbb{R}) = \{u \in H^2(S^1, \mathbb{R}^3) \mid u_2 = u_3 = 0\}$. It therefore also suffices to consider E restricted to this subspace.

Let $\rho \in \mathrm{Diff}^\infty(S^1)$ with $E(u) = E(u(\rho))$ for all $u \in H^2(S^1, R)$. Thus

(7) $$\int_{S^1} \{u(\rho)\frac{\partial}{\partial r} u(\rho)\}d\theta = \int_{S^1} \{u\frac{\partial u}{\partial r}\}d\theta \ .$$

Taking the Fréchet derivative of both sides of (7) with respect to u we can conclude that

(8) $$\int_{S^1} \{h(\rho)\frac{\partial}{\partial r} u(\rho)\}d\theta = \int_{S^1} \{h\frac{\partial u}{\partial r}\}d\theta$$

for all $u, h \in H^2(S^1, \mathbb{R})$. (8) is a restatement of the equivariance of $\langle\langle , \rangle\rangle$ under the action of $\mathcal{H}$.

Let u be the restriction to S^1 of the projection function $p_1 : \mathbb{R}^2 \to R$, $p_1(x, y) = x$. From (8) we can conclude that

(9) $$\int_{S^1} h(\rho) \frac{\partial \rho_1}{\partial r} d\theta = \int_{S^1} h \cos \theta d\theta$$

where ρ_1 denotes the first component of ρ . Let $f(\theta) = h(e^{i\theta})$ and $\rho(e^{i\theta}) = e^{i\phi(\theta)}$. Then the right side of (9) can be written as $\int_{S^1} f(\theta)\cos \theta d\theta$. Making a change of variables this is equal to

(10) $$\int_{S^1} f(\phi(\theta))\phi'(\theta)\cos \phi(\theta)d\theta = \int_{S^1} h(\rho(e^{i\theta}))\phi'(\theta)\cos \phi(\theta)d\theta \ .$$

Now

$$\frac{d}{d\theta}\rho(e^{i\theta}) = ie^{i\phi(\theta)} \cdot \phi'(\theta)$$

$$= -\phi'(\theta)\sin \phi(\theta) + i\phi'(\theta)\cos \phi(\theta) \ .$$

But

$$\frac{d}{d\theta}\rho(e^{i\theta}) = \frac{d}{d\theta}(\rho_1(e^{i\theta}), \rho_2(e^{i\theta})) .$$

Therefore

$$\frac{d}{d\theta}\rho_2(e^{i\theta}) = \phi'(\theta)\cos\phi(\theta)$$

and using (10) the right side of (9) becomes $\int_{S^1} h(\rho)\frac{\partial\rho_2}{\partial\theta}\,d\theta$. Consequently we see that

$$\int_{S^1} h(\rho)\frac{\partial\rho_1}{\partial r}\,d\theta = \int_{S^1} h(\rho)\frac{\partial\rho_2}{\partial\theta}\,d\theta$$

for all h which implies that

$$\frac{\partial\rho_1}{\partial r} = \frac{1}{r}\frac{\partial\rho_2}{\partial\theta} , \quad r = 1 \tag{11}$$

Similarly we can conclude that

$$\frac{\partial\rho_2}{\partial r} = -\frac{1}{r}\frac{\partial\rho_1}{\partial\theta} , \quad r = 1 \tag{12}$$

Equations (11) and (12) refer to the normal and tangential derivatives of the harmonic extensions of ρ_1 and ρ_2 to the disc $\mathcal{D}$ evaluated at the boundary. It is not hard to see that if (11) and (12) are satisfied for ρ_1 and ρ_2 on S^1 they are satisfied for their respective harmonic extensions $\hat{\rho}_1$ and $\hat{\rho}_2$ on all of $\mathcal{D}$. But (11) and (12) are the Cauchy-Riemann equations for the pair $\hat{\rho}_1, \hat{\rho}_2$ and hence the harmonic extension $\hat{\rho}$ of ρ is analytic and it must take the disc to itself (this requires some additional argument). We must now show that $\hat{\rho}$ is invertible. Since $\mathcal{D}$ is simply connected it suffices to show that $\hat{\rho}$ is an immersion and this will imply that $\hat{\rho}$ has an analytic inverse. Certainly $\hat{\rho}$ is an immersion on S^1 (by assumption). Suppose $\hat{\rho}$ has a singularity $p \in \mathcal{D}^0$, where $\hat{\rho}'(p) = 0$. Let $q = \hat{\rho}(p)$. Then $q \in \mathcal{D}^0$. By the analyticity of $\hat{\rho}$ this singularity must be isolated and so there is a disc $\mathcal{D}(p) \subset \mathcal{D}^0$ about p with the property that $\mathcal{D}(p)$

does not contain any other point in $\hat{\rho}^{-1}(q)$. Since $\hat{\rho}'(p) = 0$ it follows that the local degree of $\hat{\rho}$, $\deg(\hat{\rho}, \mathcal{D}(p), q) > 1$.

Let p_j, $j = 2, \ldots, m$ be the other points in $\hat{\rho}^{-1}(q)$ and set $p_1 = p$ (again by the analyticity of $\hat{\rho}$ we know there are only finitely many such p_j) . Furthermore let $\mathcal{D}_j(p_j)$ be mutually disjoint discs about each p_j . By an elementary degree theory argument one can show that

$$\deg \rho = \sum_j \deg(\hat{\rho}, \mathcal{D}(p_j), q) > 1 . \tag{13}$$

The left-hand side of this equation represents the degree of ρ as a map of S^1 to S^1 . But since ρ is an orientation preserving diffeomorphism $\deg \rho = 1$. Whence equation (13) yields a contradiction unless there are no singularities of $\hat{\rho}$ in the interior of $\mathcal{D}$. Thus $\hat{\rho}$ is an analytic diffeomorphism and $\rho \in \mathcal{H}$ comes from the restriction to S^1 of an analytic diffeomorphism of $\mathcal{D}$. Therefore $\mathcal{H} \subset \hat{G}$ and by maximality $\mathcal{H} = \hat{G}$. This concludes the proof of Theorem 7 and this section.

VIII. THE REGULARITY RESULTS FOR MINIMAL SURFACES.

In this section we state relatively recent and important results on the boundary behavior of minimal surfaces which will be crucial to our later work.

Let $u \in \overline{\mathcal{M}}_\alpha \subset \mathcal{N}_\alpha$ be a minimal surface. Recall that $\mathcal{M}_\alpha$ is the open submanifold of $\mathcal{N}_\alpha$ consisting of those $u \in \mathcal{N}_\alpha$ which map S^1 diffeomorphically onto $(S^1) = \Gamma$. The following modified regularity result is taken from Nitsche [17]. It is an improvement of the first such result due to Hildebrandt [13].

THEOREM 1. Let Γ be the image of a $C^{m,\rho}$ embedding $\alpha : S^1 \to \mathbb{R}^3$, where $C^{m,\rho}$, $m \geq 1$, $0 < \rho < 1$ is the Hölder space of m times continuously differentiable maps with Hölder exponent ρ. Then the harmonic extension of any critical point $u \in \overline{\mathcal{M}}_\alpha$ of E_α is also a $C^{m,\rho}$ map of $\mathcal{D}$ to $\mathbb{R}^3$, $\mathcal{D}$ the closed disc.

Fix three points P_1, P_2, and P_3 on S^1. Let $\mathcal{S}_\alpha \subset \mathcal{N}_\alpha$ be the submanifold of maps u with $u(P_i) = \alpha(P_i)$. Then the $C^{m,\rho}$ norm on $\mathcal{D}$ of any critical point u of E_α in $\mathcal{S}_\alpha \cap \overline{\mathcal{M}}_\alpha$ can be chosen independently of u.

COROLLARY. Let Γ be the image of a C^{k+2} embedding α, with $k \geq 2$. Then the set of critical points of E_α in $\mathcal{S}_\alpha \cap \overline{\mathcal{M}}_\alpha$ is compact.

<u>Proof</u> (of corollary). Since $C^{k+1}(S^1, \mathbb{R}^3) \subset C^{3,\rho}(S^1, \mathbb{R}^3)$, $0 < \rho \leq 1/2$, it follows that Γ is of class $C^{3,\rho}$ for any ρ. Thus by Theorem 1 any critical point in $\mathcal{S}_\alpha \cap \overline{\mathcal{M}}_\alpha$ considered as a harmonic map on the closed disc is uniformly bounded in the $C^{3,\rho}$ norm. The inclusion $C^{3,\rho}(\mathcal{D}, \mathbb{R}^3) \longrightarrow H^3(\mathcal{D}, \mathbb{R}^3)$ and the restriction map $H^3(\mathcal{D}, \mathbb{R}^3) \longrightarrow H^{5/2}(S^1, \mathbb{R}^3)$ are continuous. Therefore any $u \in \mathcal{S}_\alpha \cap \overline{\mathcal{M}}_\alpha$ which is a critical point of E_α is a critical point of E is uniformly bounded in $H^{5/2}(S^1, \mathbb{R}^3)$ norm. Since the inclusion of $H^{5/2}(S^1, \mathbb{R}^3)$ to $H^2(S^1, \mathbb{R}^3)$ is completely continuous the result follows.

Another version of Theorem 1 has also been proven by Heinz and Tomi [12].

Let $\omega = x + iy$ denote the complex coordinate system for $\mathbb{R}^2$. If $u \in \mathcal{N}_\alpha$ is a minimal surface then the component functions of u appear as the real parts of analytic functions on $\mathcal{D}^0$, $u_1(x,y) = \text{Re } f_1(\omega)$, $u_2(x,y) = \text{Re } f_2(\omega)$, and $u_3(x,y) = \text{Re } f_3(\omega)$, for which Monges' relation $[f_1'(\omega)]^2 + [f_2'(\omega)]^2 + [f_3'(\omega)]^2 = 0$ holds.

DEFINITION. Let $u \in \mathcal{N}_\alpha$ be a critical point of E_α. A point $p \in$ is said to be a <u>branch point</u> of u if the harmonic extension $\hat{u}$ of u to $\mathcal{D}$ fails to be an immersion at p.

Despite important recent work on branch points due to Osserman [9] and Gulliver [10] and Alt, little is known about the nature of an embedding α which forces the existence of branch points in those minimal surfaces which are not relative minima.

The next interesting result concerning boundary branch points is due to Nitsche [17]. The corresponding result for interior branch points is trivial.

THEROEM 2. If Γ is a regular Jordan curve of class $C^{2,\rho}$, $0 < \rho < 1$ and if $\omega_1 = e^{i\theta_1}$ is a boundary branch point for a classical simply connected minimal surface u, then for ω close enough to ω_1 the asymptotic expansion

$$u_x(\omega) - i\,u_y(\omega) = \{a, b, c\}\ (\omega - \omega_1)^{2p} + 0(|\omega - \omega_1|^{2p})$$

is valid. Here a, b, c are three complex numbers satisfying the conditions $a^2 + b^2 + c^2 = 0$, $|a| + |b| + |c| > 0$, with p a positive integer.

As an important consequence we have

COROLLARY. Any critical point $u \in \overline{\mathcal{M}}_\alpha$ of E_α can have at most a finite number of branch points.

§IX. THE FRÉCHET DERIVATIVE OF THE MINIMAL SURFACE VECTOR X AND THE SURFACE FIBRE BUNDLE

In §III we introduced the space $H^{r+2}, \mathbb{R}^3)$ and the open subset $\mathcal{A} \subset H^{r+2}(S^1, \mathbb{R}^3)$ consisting of embeddings. Given a fixed $\alpha \in \mathcal{A}$ we formed the C^r submanifold $\mathcal{N}_\alpha$ of $H^2(S^1, \mathbb{R}^3)$. By harmonic extension $\mathcal{N}_\alpha$ can be thought of as the manifold of harmonic surfaces spanning $\alpha(S^1)$. Let $\mathcal{N} = {}_\alpha\bigcup \mathcal{N}_\alpha$ and $\pi : \mathcal{N} \longrightarrow \mathcal{A}$ be the natural projection map $\pi(\mathcal{N}_\alpha) = \alpha$. The space $\mathcal{N}$ (as we shall see in detail later on) has the structure of a C^r fibre bundle over the C^∞ Hilbert manifold $\mathcal{A}$. It is the fibre bundle of harmonic surfaces. Let $T\mathcal{N}$ denote the tangent bundle of $\mathcal{N}$ with $\pi_0 : T\mathcal{N} \longrightarrow \mathcal{N}$ the tangent bundle projection map. For each $n \in \mathcal{N}$, $\pi(n) = \alpha$ the space tangent to $\mathcal{N}_\alpha$ at n, $T_n\mathcal{N}_\alpha$ is a closed subspace of $T_n\mathcal{N}$; in fact $T_n\mathcal{N}_\alpha = \operatorname{Ker} \pi_*(n)$, the kernel of the Fréchet derivative π_* of π evaluated at n.

$\pi_0 : \bigcup_n T_n\mathcal{N}_{\pi(n)} \longrightarrow \mathcal{N}$ is a subbundle of the tangent bundle $T\mathcal{N}$ which we shall call the <u>vertical subbundle</u> of $T\mathcal{N}$ induced by the fibering $\pi : \mathcal{N} \longrightarrow \mathcal{A}$ and we shall denote it by $v(T\mathcal{N})$.

For each $\alpha \in \mathcal{A}$ we have a vector field X^α whose zeros are the simply connected minimal surfaces spanning $\alpha(S^1)$. We introduce a <u>new</u> C^{r-1} vector field X on $\mathcal{N}$ defined by $X(n) = X^\alpha(n)$ if $\pi(n) = \alpha$. It follows immediately from its definition that X is a section of the vertical subbundle $v(T\mathcal{N})$. We call such a vector field on a fibre bundle $\pi : \mathcal{N} \longrightarrow \mathcal{A}$ a <u>vertical</u> vector field (cf. §I). Such vector fields play a central role in intrinsically non-linear problems in analysis.

Our two main goals in this section are first to compute the Fréchet derivative $X_*^\alpha(u) : T_u\mathcal{N}_\alpha \longrightarrow T_u\mathcal{N}_\alpha$ at a zero $u \in \mathcal{N}_\alpha$ of X^α. This derivative coincides with the covariant derivative of X^α at u (cf. remarks of §I). Second, we wish to show explicitly that $\mathcal{N}$ is a fibre bundle over $\mathcal{A}$ and to compute the Fréchet derivative $X_*(u)$, of the vector field X on $\mathcal{N}$ at a zero u. Since X is vertical the linear map $X_*(u)$ maps $T_u\mathcal{N}$ into the subspace $T_u\mathcal{N}_{\pi(u)}$ of $T_u\mathcal{N}$.

To start on this program we note that each $\alpha \in \mathcal{A}$ determines the map $\Omega : \alpha(S^1) \longrightarrow \mathcal{OP}(\mathbb{R}^3)$. Again in order to keep track of the embedding α <u>we shall henceforth denote this map by</u> Ω^α. It is clear that if two embeddings α and ρ have the same image then $\Omega^\alpha = \Omega^\rho$. Recall that

$$\Omega^\alpha(u)\frac{\partial X^\alpha}{\partial r}(u) = \Omega^\alpha(u)\frac{\partial u}{\partial r} \tag{1}$$

is the defining equation for $X^\alpha : \mathcal{N}_\alpha \longrightarrow T\mathcal{N}_\alpha$. If $k \in H^2(S^1, \mathbb{R}^3)$, then we have an explicit formula for $\Omega^\alpha(u)k$, namely

$$(\Omega^\alpha(u)k)(\theta) = \left\langle \frac{\alpha'(\alpha^{-1}(u(\theta)))}{\|\alpha'(\alpha^{-1}(u(\theta)))\|}, k(\theta) \right\rangle \frac{\alpha'(\alpha^{-1}(u(\theta)))}{\|\alpha'(\alpha^{-1}(u(\theta)))\|} \tag{2}$$

where $\langle , \rangle$ represents the $\mathbb{R}^3$ inner product, $\| \ \|$ the Euclidean norm in $\mathbb{R}^3$, and $\alpha'(\alpha^{-1}(u(\theta)))$ the oriented unit tangent vector to $\alpha(S^1)$ at the point $u(\theta)$ ($\alpha'(s)$ denotes the derivative $\frac{d}{ds}\alpha(s)$).

If u is a zero of X^α in $\mathcal{M}_\alpha$ then by Theorem 2, §VIII, u can have at most finitely many singularities or branch points on S^1; i.e., $u'(\theta) = 0$ for at most finitely many θ. In this case formula (2) transforms to

$$(\Omega^\alpha(u)k)(\theta) = \left\langle \frac{u_\theta}{\|u_\theta\|}, k \right\rangle \frac{u_\theta}{\|u_\theta\|} \tag{2'}$$

where (2') holds for all but finitely many θ.

Returning to equation (1), taking the Fréchet derivative of both sides at zero u of X^α, and noting that the order of differentiation is irrelevant, we get that for $h \in T_u \mathcal{N}_\alpha$

$$\Omega^\alpha(u)\frac{\partial}{\partial r}[X_*^\alpha(u)h] = \Omega_*^\alpha(u)[h][\frac{\partial u}{\partial r}] + \Omega^\alpha(u)\frac{\partial h}{\partial r} \tag{3}$$

where $\Omega_*^\alpha(u)$ is the Fréchet derivative at u. Recall that if u is a zero of X^α then $\left\langle \frac{\partial u}{\partial \theta}, \frac{\partial u}{\partial r}\right\rangle = 0$. From (2') it follows that for $u \in \overline{\mathcal{M}}_\alpha$

$$\Omega^\alpha(u)\frac{\partial}{\partial r}[X_*^\alpha(u)h] = \left\langle \frac{h_\theta}{\|u_\theta\|}, u_r\right\rangle \frac{u_\theta}{\|u_\theta\|} + \left\langle \frac{u_\theta}{\|u_\theta\|}, h_r\right\rangle \frac{u_\theta}{\|u_\theta\|}. \tag{4}$$

The energy functional $E_\alpha : \mathcal{N}_\alpha \to R$ has Fréchet derivative $DE_\alpha(u)[k] = \int_{S^1} \left\langle \frac{\partial X^\alpha}{\partial r}(u), k\right\rangle d\theta$. If u is a zero of X^α (and hence a critical point of E_α) the Hessian of E_α at u is given by

$$D^2E_\alpha(u)(h, k) = \int_{S^1} \left\langle \frac{\partial}{\partial r} X_*^\alpha(u)[h], k\right\rangle d\theta .$$

Using (4) we see that this is equal to

$$= \int_{S^1} \left\langle \Omega^\alpha(u)\frac{\partial}{\partial r}[X_*^\alpha(u)h], k\right\rangle d\theta$$

$$= \int_{S^1} \left\langle \left\langle \frac{h_\theta}{\|u_\theta\|}, u_r\right\rangle \frac{u_\theta}{\|u_\theta\|} + \left\langle \frac{u_\theta}{\|u_\theta\|}, h_r\right\rangle \frac{u_\theta}{\|u_\theta\|}, k\right\rangle d\theta$$

$$= \int_{S^1} \left\langle \frac{h_\theta}{\|u_\theta\|}, u_r\right\rangle \left\langle \frac{u_\theta}{\|u_\theta\|}, k\right\rangle d\theta + \int_{S^1} \left\langle \frac{\partial h}{\partial r}, k\right\rangle d\theta .$$

Using the fact that $\|u_r\| = \|u_\theta\|$ we find this equal to

$$\int_{S^1} \left\langle h_\theta, \left\langle k, \frac{u_\theta}{\|u_\theta\|}\right\rangle \frac{u_r}{\|u_r\|}\right\rangle d\theta + \int_{S^1} \left\langle \frac{\partial k}{\partial r}, h\right\rangle d\theta .$$

Letting $U(\theta) = \frac{u_\theta}{\|u_\theta\|}$, the "unit tangent", and $N(\theta) = \frac{u_r}{\|u_r\|}$, the "unit normal", and integrating by parts we obtain the important and fundamental Hessian formaula for E_α at a critical point $u \in \overline{\mathcal{M}}_\alpha$:

(5) $$D^2E_\alpha(u)(h), k) = \int_{S^1} \left\langle X_*^\alpha(u)h, \frac{\partial k}{\partial r} \right\rangle d\theta$$

$$= \int_{S^1} \left\langle \frac{\partial k}{\partial r}, h \right\rangle d\theta - \int_{S^1} \left\langle \frac{d}{d\theta}[N \left\langle k, U \right\rangle, h \right\rangle d\theta \; .$$

Using the fact that the Hessian is a symmetric bilinear form we can further conclude that

(6) $$\Omega^\alpha(u)\frac{\partial}{\partial r}X_*^\alpha(u)k = \Omega^\alpha(u)[\frac{\partial k}{\partial r} - \frac{d}{d\theta}(N \left\langle k, U \right\rangle)] \; .$$

Again recalling that $h \longmapsto L_u h = \Omega^\alpha(u)\frac{\partial h}{\partial r}$ is an isomorphism from $T_u\mathcal{N}_\alpha$ to $(T_u\mathcal{N}_\alpha)^1$ we obtain the fundamental formula for the Fréchet derivative of the minimal surface vector field X^α, namely

(7) $$X_*^\alpha(u)k = L_u^{-1} \left\{\Omega^\alpha(u)[\frac{\partial k}{\partial r} - \frac{d}{d\theta}(N \left\langle k, U \right\rangle)]\right\} \; .$$

Since the vector field X^α is Palais-Smale it follows that the Hessian of E_α given in formula (5) has finite index and nullity. However, in the case $u \in \mathcal{M}_\alpha$ ($u_\theta \neq 0$, for all θ) we can easily verify this directly from formula (5). We wish to show that the linear subspace $\mathcal{F} \subset T_u\mathcal{N}_\alpha$ with $D^2E_\alpha(u)(k, k) \leq 0$ for all $k \in \mathcal{F}$ is finite dimensional. From formula (5) it is evident that for all $k \in \mathcal{F}$

$$\int_{S^1} \left\langle \frac{\partial k}{\partial r}, k \right\rangle d\theta - \int_{S^1} \left\langle \frac{d}{d\theta}[N \left\langle k, U \right\rangle], k \right\rangle d\theta \leq 0 \; .$$

Noting that $N(\theta)$ is normal to $\alpha(S^1)$ at $u(\theta)$ we see that this inequality is equivalent to

$$\int_{S^1} \left\langle \frac{\partial k}{\partial r}, k \right\rangle d\theta - \int_{S^1} \left\langle \left\langle k, U \right\rangle N', k \right\rangle d\theta \leq 0$$

where $N' = \frac{dN}{d\theta}$. (By the regularity Theorem 1, §VIII, $\theta \longrightarrow N'(\theta)$ exists and is continuous.) Therefore, for all $k \in \mathcal{F} \subset T_u\mathcal{N}_\alpha$

(8) $$\int_{S^1} \left\langle \frac{\partial k}{\partial r}, k \right\rangle d\theta \leq \int_{S^1} \left\langle N', k \right\rangle \left\langle k, U \right\rangle d\theta$$

$$\leq C \int_{S^1} |k|^2 d \; = C \, \|k\|_{H^0}$$

where C is some constant. From the lemma to Theorem 2, §IV, we know that the triple norm $|||\, k \,|||^2 = \int_{S^1} \left\langle \frac{\partial k}{\partial r}, k \right\rangle$ on $T_u \mathcal{N}_\alpha$ is equivalent to the $H^{1/2}$ norm. Thus inequality (8) states that for all $k \in \mathcal{F}$

$$\|k\|_{H^{1/2}} \leq C \, \|k\|_{H^0} \; .$$

Since the inclusion of $H^{1/2}$ into H^0 is completely continuous this implies that the unit ball of $\mathcal{F}$ is compact which implies that $\mathcal{F}$ is finite dimensional, and hence that the index and nullity of the Hessian of E_α at u is finite.

We know from §VII that X^α is equivariant with respect to the action of the conformal group G. This implies that at a zero u of X^α there is a "submanifold" of zeros of X^α through u, namely the orbit $\mathcal{O}_G(u)$ of u under G. Thus $T_u \mathcal{O}_G(u)$, the tangent space to $\mathcal{O}_G(u)$ at u, is always in the kernel, $\operatorname{Ker} X^\alpha_*(u)$, of the Fréchet derivative of X^α at u. Therefore, the dimension of this kernel is at least three. From the corollary to Theorem 2, §VII, we know that $T_u \mathcal{O}_G(u)$ is spanned by u_θ, $\sin \theta u_\theta$, and $\cos \theta u_\theta$. A direct computation shows that if k is in the subspace spanned by these vectors then $X^\alpha_*(u) k = 0$ which by (7) means that

$$\Omega^\alpha(u) \frac{\partial k}{\partial r} = \Omega^\alpha(u) [\langle k, U \rangle N'] \; .$$

We shall, however, omit this computation, and proceed to the computation of the Fréchet derivative $X_*(u)$ of the vertical vector field $X : \mathcal{N} \rightarrow T\mathcal{N}$.

Let $\alpha_0 \in \mathcal{A}$ be a fixed embedding, $\pi : \mathcal{N} \rightarrow \mathcal{A}$ the surface fibre bundle introduced at the beginning of this section, $\mathcal{W} \subset H^{r+2}(S^1, \mathbb{R}^3)$ a sufficiently small neighborhood of 0 with the property that $\alpha_0 + h \in \mathcal{A}$

for all $h \in \mathcal{W}$, and finally $\mathcal{W}_{\alpha_0}$ the neighborhood of α_0 given by $\alpha_0 + \mathcal{W}$. Then we can trivialize $\pi^{-1}(\mathcal{W}_{\alpha_0})$ as follows. For all $u \in \mathcal{N}_{\alpha_0}$ we define the fibre preserving diffeomorphism

$$\Xi : \mathcal{W}_{\alpha_0} \times \mathcal{N}_{\alpha_0} \longrightarrow \pi^{-1}(\mathcal{W}_{\alpha_0})$$

by $\Xi(\rho, u) = \rho(\alpha_0^{-1}(u))$, where $\rho = \alpha_0 + h$, $h \in \mathcal{W}$.

Define the C^r map $\Omega : \mathcal{N} \to \mathcal{L}(H^2, H^2)$, where $\mathcal{L}^2(H^2, H^2)$ denotes the linear maps of $H^2(S^1, \mathbb{R}^3)$ to itself, by

$$\Omega(v)[\xi] = \left\langle \frac{\rho'(\rho^{-1}(u_\rho))}{\|\rho'(\rho^{-1}(u_\rho))\|}, \xi \right\rangle \frac{\rho'(\rho^{-1}(u_\rho))}{\|\rho'(\rho^{-1}(u_\rho))\|}$$

where $\xi \in H^2(S^1, \mathbb{R}^3)$ and $v = u_\rho \in \mathcal{N}$.

For each $\rho \in \mathcal{A}$, $\mathcal{N}_\rho$ is a submanifold of $H^2(S^1, \mathbb{R}^3)$. The fibre bundle $\mathcal{N}$ can be embedded as a submanifold of the product space $\mathcal{A} \times H^2(S^1, \mathbb{R}^3)$ and therefore as a submanifold of $H^{r+2}(S^1, \mathbb{R}^3) \times H^2(S^1, \mathbb{R}^3)$ by sending $u_\rho \in \mathcal{N}_\rho$ to the point (ρ, u_ρ), and we shall henceforth think of $\mathcal{N}$ as a submanifold of this product space via the prescribed embedding. Doing this we can define a new trivialization Φ of $\pi^{-1}(\mathcal{W}_{\alpha_0})$,

$$\Phi : \mathcal{W}_{\alpha_0} \times \mathcal{N}_{\alpha_0} \longrightarrow \subset \mathcal{A} \times H^2$$

by $\Phi(\rho, u) = (\rho, \rho(\alpha_0^{-1}(u)))$. The Fréchet derivative of Φ at the point (α_0, u) operating on the tangent vector $(h, k) \in H^{r+2} \times H^2$ is given by

(9) $$\Phi_*(\alpha_0, u)[h, k] = D\Phi(\rho, u)[h, k] = (h, h(\alpha_0^{-1}(u)) + k) .$$

Also Ω at the point $\Phi(\rho, u)$ operating on $\xi \in H^2$ is given by

(10) $$\Omega(\Phi(\rho, u))[\xi] = \left\langle \frac{\rho'(\alpha_0^{-1}(u))}{\|\rho'(\alpha_0^{-1}(u))\|}, \xi \right\rangle \frac{\rho'(\alpha_0^{-1}(u))}{\|\rho'(\alpha_0^{-1}(u))\|} .$$

Let $u \in \mathcal{N}_{\alpha_0}$ be a zero of X. Then $\Phi(\alpha_0, u) = (\alpha_0, u)$ and using formulas (9) and (10) we obtain a formula for the Fréchet derivative of the composition $(\rho, w) \rightsquigarrow \Omega(\Phi(\rho, w))[\frac{\partial u}{\partial r}]$ the point (α_0, u) applied to the vector $(h, k) \in H^{r+2} \times H^2$, namely

$$(\Omega_*(\Phi(\alpha_0, u)) \circ \Phi_*(\alpha_0, u)[h, k])[\frac{\partial u}{\partial r}] = \tag{11}$$

$$\left\langle \frac{h'(\alpha_0^{-1}(u))}{\|\alpha_0'(\alpha_0^{-1}(u))\|} + \frac{k'(\theta)}{\|u_\theta\|}, \frac{\partial u}{\partial r} \right\rangle \frac{\alpha'_0(\alpha_0^{-1}(u))}{\|\alpha'_0(\alpha_0^{-1}(u))\|} .$$

Let $\pi_2 : \mathcal{A} \times H^2(S^1, \mathbb{R}^3) \rightarrow H^2(S^1, \mathbb{R}^3)$ be the projection onto the second factor and let $\bar{\pi}_2$ be the restriction of π_2 to $\mathcal{N}$. Then, locally, we find that the defining partial differential equation for $X : \mathcal{N} \rightarrow T\mathcal{N}$ is

$$\Omega(\Phi(\rho, w))\frac{\partial}{\partial r} X(\Phi(\rho, w)) = \Omega(\Phi(\rho, w))\frac{\partial}{\partial r} \bar{\pi}_2\Phi(\rho, w) . \tag{12}$$

Taking the Fréchet derivative of (12) at a zero $u \in \mathcal{N}_{\alpha_0}$ and identifying u and $(\alpha_0, u) = \Phi(\alpha_0, u)$ we obtain

$$\Omega(u)\frac{\partial}{\partial r} X_*(u)(\Phi_*(\alpha_0, u)[h, k]) = (\Omega_*(u) \circ \Phi_*(\alpha_0, u)[h, k])[\frac{\partial u}{\partial r}] +$$

$$\Omega(u)[\frac{\partial}{\partial r} \pi_2 \Phi_*(\alpha_0, u)[h, k]] = (\Omega_*(u) \circ \Phi_*(\alpha_0, u)[h, k])[\frac{\partial u}{\partial r}] +$$

$$\left\langle \frac{\alpha'_0(\alpha_0^{-1}(u))}{\|\alpha'_0(\alpha_0^{-1}(u))\|}, \frac{\partial}{\partial r} h(\alpha_0^{-1}(u)) \right\rangle \frac{\alpha'_0(\alpha_0^{-1}(u))}{\|\alpha'_0(\alpha_0^{-1}(u))\|} + \left\langle \frac{\alpha'_0(\alpha_0^{-1}(u))}{\|\alpha'_0(\alpha_0^{-1}(u))\|}, \frac{\partial k}{\partial r} \right\rangle \frac{\alpha'_0(\alpha_0^{-1}(u))}{\|\alpha'_0(\alpha_0^{-1}(u))\|} .$$

For $u \in \mathcal{M}_{\alpha_0}$ this simplifies to

$$(\Omega_*(u) \circ \Phi_*(\alpha_0, u)[h, k])[\frac{\partial u}{\partial r}] + \left\langle \frac{u_\theta}{\|u_\theta\|}, \frac{\partial}{\partial r} h(\alpha_0^{-1}(u)) + \frac{\partial k}{\partial r} \right\rangle \frac{u_\theta}{\|u_\theta\|}$$

which by (11) equals

$$\left\langle \frac{d}{d\theta}(h(\alpha_0^{-1}(u)) + k), \frac{u_r}{\|u_r\|} \right\rangle \frac{u_\theta}{\|u_\theta\|} \tag{13}$$

$$+ \left\langle \frac{u_\theta}{\|u_\theta\|}, \frac{\partial}{\partial r} h(\alpha_0^{-1}(u)) + \frac{\partial k}{\partial r} \right\rangle \frac{u_\theta}{\|u_\theta\|} = T(h, k) .$$

In the above derivation we have used the fact that $\|u_\theta\| = \|u_r\|$. Thus

$$T(h, k) = \Omega(u)\frac{\partial}{\partial r} X_*(u)(\Phi_*(\alpha_0, u)[h, k])$$

an important expression we shall use in §XIII.

We end this section by noting that geometrically the Plateau problem does not depend on the particular parameterization α of a curve $\Gamma \subset \mathbb{R}^3$. This is reflected in our setup as follows. If α and β are parameterizations of the same curve, then geometrically the manifolds $\mathcal{N}_\alpha$ and $\mathcal{N}_\beta$ coincide as submanifolds of $H^2(S^1, \mathbb{R}^3)$. Therefore an element $u \in \mathcal{N}_\alpha$ can simultaneously be viewed as an element of $\mathcal{N}_\beta$ by identifying these two spaces. It follows from the defining equations for X^α and X^β that $X^\alpha(u) = X^\beta(u)$ for all $u \in \mathcal{N}_\alpha = \mathcal{N}_\beta$.

Therefore the vector fields $\{X^\alpha\}_{\alpha\in\mathcal{A}}$ truly depend on the image of the embeddings α , rather than the embedding itself.

§X. THE MINIMAL SURFACE VECTOR FIELD X IS PROPER ON BOUNDED SETS

In [20] and [28] Palais and Smale introduced the notion of condition (C) for real valued maps on Hilbert or Banach manifolds. This condition was essential for Palais and Smale's development of Morse theory on Hilbert manifolds and for Palais' development of Lusternik-Schnirelman theory on Banach manifolds. In [32] the author showed that condition (C) was not essential for Morse theory on either Hilbert or Banach manifolds. In this paper we studied "gradient like" vector fields which satisfied condition (CV) and showed that the existence of such vector fields was sufficient for Morse theory. In §I of this paper we introduced the notion of proper vector field on a Finsler manifold. A vector field on a Finsler manifold $\mathcal{M}$ satisfies condition (CV) if it is proper on bounded subsets of $\mathcal{M}$. For the sake of completeness let us review a few definitions.

DEFINITION. A C^r Banach Finsler manifold $\mathcal{M}$ is a C^r smooth Banach manifold together with a function $\| \ \|: T\mathcal{M} \to R$ such that

(i) $\| \ \|_x : T_x\mathcal{M} \to R$ is an admissible norm

(ii) $x \mapsto \| \ \|_x$ is continuous.

A Finsler structure on $\mathcal{M}$ determines a metric d on $\mathcal{M}$ which induces a topology compatible with the given topology. For more details consult [21].

If $\mathcal{N} \subset \mathcal{M}$ is a submanifold then the Finsler on $\mathcal{M}$ induces a Finsler on $\mathcal{N}$; in particular if $\mathcal{N} \subset H$ is a submanifold of a Banach space H then the norm on H induces a Finsler on $\mathcal{N}$.

DEFINITION. Let $\mathcal{M}$ be a Banach Finsler manifold. A set $\mathcal{S} \subset \mathcal{M}$ is <u>bounded</u> if $\sup_{p,q \in \mathcal{S}} d(p, q) < \infty$.

DEFINITION. A vector field $X : \mathcal{M} \to T\mathcal{M}$ on a Finsler manifold $\mathcal{M}$ satisfies <u>condition</u> (CV), or equivalently X is proper on bounded sets, if whenever $\|X(p_j)\| \to 0$ for a bounded sequence $\{p_j\}$, then there is a subsequence of $\{p_j\}$ which converges.

The first goal of this section is to show that $X^\alpha : \mathcal{N}_\alpha \to T\mathcal{N}_\alpha$ satisfies condition (CV). The manifold $\mathcal{N}_\alpha$ is a submanifold of $H^2(S^1, \mathbb{R}^3)$ and the norm on H^2 induces a Finsler on $\mathcal{N}_\alpha$. A set is bounded in $\mathcal{N}_\alpha$ iff it is bounded in $H^2(S^1, \mathbb{R}^3)$ and we shall use this fact.

THEOREM 1. $X^\alpha : \mathcal{N}_\alpha \to T\mathcal{N}_\alpha$ satisfies condition (CV).

Proof. Let $\{u_n\}$ be a bounded sequence in $\mathcal{N}_\alpha$ (and hence in $H^2(S^1, \mathbb{R}^3)$) with $\|X^\alpha(u_n)\|_{H^2(S^1, \mathbb{R}^3)} \to 0$ as $n \to \infty$. We must show that $\{u_n\}$ has a convergent subsequence.

Henceforth, for notational convenience we shall denote the $H^s(S^1, \mathbb{R}^m)$ norm by $\|\ \|_{H^s}$ and the $H^s(\mathcal{D}, \mathbb{R}^m)$ norm by $\|\ \|_{H^s_{\mathcal{D}}}$. Recall, once again, that for $u \in \mathcal{N}_\alpha$, X^α satisfies the differential relation $\Omega^\alpha(u)\frac{\partial X^\alpha}{\partial r}(u) = \Omega^\alpha(u)\frac{\partial u}{\partial r}$. If $\|X^\alpha(u_n)\|_{H^2} \to 0$ then $\|\frac{\partial}{\partial r} X^\alpha(u_n)\|_{H^1} \to 0$ and consequently

$$\Omega^\alpha(u_n)\frac{\partial X^\alpha}{\partial r}(u_n) \to 0 \quad \text{and} \quad \Omega^\alpha(u_n)\frac{\partial u_n}{\partial r} \to 0 \tag{1}$$

where the convergence of these sequences is in the $H^1(S^1, \mathbb{R}^3)$ norm.

As in §IV let $\hat{\Omega}^\alpha(u_n)$ be the harmonic extension of $\hat{\Omega}^\alpha(u_n)$ to $\mathcal{D}$ and for $h : S^1 \to \mathbb{R}^3$ let $\hat{h} : \mathcal{D} \to \mathbb{R}^3$ denote its harmonic extension. From the methods of Theorems 4 and 5 of §IV it follows that if $\{h_n\}$ is a sequence in $H^1(S^1, \mathbb{R}^3)$ with $h_n(\theta) \in T_{u_n(\theta)}\Gamma$ for all θ then

$$\hat{\Omega}^\alpha(u_n)\hat{h}_n = \hat{h}_n + \psi_n \tag{2}$$

where $\psi_n \in H^{5/2}(\mathcal{D}, \mathbb{R}^3)^0$. Thus $\frac{\partial \psi_n}{\partial r} \in H^{3/2}(\mathcal{D}, \mathbb{R}^3)$ and $\frac{\partial \psi_n}{\partial r}\Big|S^1 \in H^1(S^1, \mathbb{R}^3)$. The map ψ_n satisfies the elliptic equation

$$(3) \qquad \Delta\psi_n = 2\nabla\hat{\Omega}^\alpha(u_n) \cdot \nabla\hat{h}_n .$$

For a review of our notational conventions see §IV. Briefly $\Omega^\alpha(u)$ is represented by a 3×3 matrix $a_{jk}(u) \in H^2(S^1, \mathbb{R}^3)$ and $\nabla\hat{\Omega}^\alpha(u_n) \cdot \nabla\hat{h}_n \in H^{1/2}(\mathcal{D}, \mathbb{R}^3)$ is the map whose ℓ^{th} column is $\sum_j \nabla\hat{a}_{\ell j}(u_n) \cdot \nabla(\hat{h}_n)_j$. Since $u_n \in H^2(S^1, \mathbb{R}^3)$ each $\nabla\hat{a}_{\ell j}(u_n) \in H^{3/2}(\mathcal{D}, \mathbb{R}^2)$.

LEMMA 1. $\|\nabla\hat{a}_{\ell j}(u)\|_{H^{3/2}_{\mathcal{D}}} \leq \text{const}\|u\|_{H^2}$ where the constant can be selected independently of u.

Proof. By definition $\Delta\hat{a}_{\ell j}(u) = 0$. From a standard regularity result ([1], page 704) in the theory of elliptic P.D.E. we get

$$(4) \qquad \|\hat{a}_{\ell j}(u)\|_{H^{5/2}_{\mathcal{D}}} \leq \text{const}\|a_{\ell j}(u)\|_{H^2} .$$

Since Γ is compact it follows that the right-hand side of (4) is bounded by $\text{const}\|u\|_{H^2}$ with the constant chosen independently of u. We also know that

$$\|\nabla\hat{a}_{\ell j}(u)\|_{H^{3/2}_{\mathcal{D}}} \leq \|\hat{a}_{\ell j}(u)\|_{H^{5/2}_{\mathcal{D}}} .$$

Putting this inequality together with the others we see that the lemma is proved.

LEMMA 2. In equation (2)

$$(5) \qquad \|\psi_n\|_{H^{5/2}_{\mathcal{D}}} \leq \text{const}\|u_n\|_{H^2} \cdot \|\hat{h}_n\|_{H^{3/2}_{\mathcal{D}}} .$$

Proof. It follows from Theorem 3, §II, that

$$\|\nabla\hat{\Omega}^{\alpha}(u_n)\cdot\nabla\hat{h}_n\|_{H^{1/2}_{\mathcal{B}}} \le \text{const}\|\nabla\hat{\Omega}^{\alpha}(u_n)\|_{H^{3/2}}\cdot\|\nabla\hat{h}_n\|_{H^{1/2}}\ .$$

Applying Lemma 1 we find the term on the right bounded by

$$\text{const}\|u_n\|_{H^2}\cdot\|\hat{h}_n\|_{H^{3/2}_{\mathcal{B}}}\ .$$

Considering the defining equation (3) for ψ_n and again applying the regularity results of [1] we get that

$$\|\psi_n\|_{H^{5/2}_{\mathcal{B}}} \le \text{const}\|u_n\|_{H^2}\cdot\|\hat{h}_n\|_{H^{3/2}_{\mathcal{B}}}$$

which completes Lemma 2.

Putting these inequalities together we see that the proof of Lemma 2 is complete.

Let us proceed with the proof of Theorem 1. Since $u_n : S^1 \to \Gamma$ and $\frac{\partial u_n}{\partial\theta} \in T_{u_n(\theta)}\Gamma$ for all θ it follows that

$$\Omega(u_n)\frac{\partial u_n}{\partial\theta} = \frac{\partial u_n}{\partial\theta}\ . \tag{6}$$

Considering equation (2) with $h_n = \frac{\partial u_n}{\partial\theta}$ and applying Lemmas 1 and 2 we see that

$$\hat{\Omega}^{\alpha}(u_n)\frac{\partial\hat{u}_n}{\partial\theta} = \frac{\partial\hat{u}_n}{\partial\theta} + \psi_n \tag{7}$$

with

$$\|\psi_n\|_{H^{5/2}_{\mathcal{B}}} \le \text{const}\|u_n\|_{H^2}\left\|\frac{\partial\hat{u}_n}{\partial\theta}\right\|_{H^{3/2}_{\mathcal{B}}} \tag{8}$$

$$\le \text{const}\|u_n\|_{H^2}\left\|\frac{\partial u_n}{\partial\theta}\right\|_{H^1}\ .$$

Differentiating both sides of (6) by $\frac{\partial}{\partial r}$ we obtain

$$\frac{\partial\hat{\Omega}^\alpha}{\partial r}(u_n)\left[\frac{\partial\hat{u}_n}{\partial\theta}\right] + \hat{\Omega}^\alpha(u_n)\frac{\partial^2\hat{u}_n}{\partial r\partial\theta} = \frac{\partial^2\hat{u}_n}{\partial r\partial\theta} + \frac{\partial\psi_n}{\partial r} .$$

But

$$\hat{\Omega}^\alpha(u_n)\frac{\partial^2\hat{u}_n}{\partial r\partial\theta} = \hat{\Omega}^\alpha(u_n)\frac{\partial}{\partial\theta}\frac{\partial\hat{u}_n}{\partial r}$$

$$= \frac{\partial}{\partial\theta}\left(\hat{\Omega}^\alpha(u_n)\frac{\partial\hat{u}_n}{\partial r}\right) - \frac{\partial\Omega^\alpha}{\partial\theta}(u_n)\left[\frac{\partial\hat{u}_n}{\partial r}\right] .$$

Therefore

(9) $$\frac{\partial^2\hat{u}_n}{\partial r\partial\theta} = \frac{\partial\hat{\Omega}^\alpha}{\partial r}(u_n)\left[\frac{\partial\hat{u}_n}{\partial\theta}\right] - \frac{\partial\hat{\Omega}}{\partial\theta}(u_n)\left[\frac{\partial\hat{u}_n}{\partial r}\right] + \frac{\partial}{\partial\theta}\left(\hat{\Omega}^\alpha(u_n)\frac{\partial\hat{u}_n}{\partial r}\right) - \frac{\partial\psi_n}{\partial r} .$$

Applying (8) we get

(10) $$\left\|\frac{\partial\psi_n}{\partial r}\right\|_{H^{3/2}_{\mathcal{D}}} \leq \text{const}\|u_n\|_{H^2} \cdot \left\|\frac{\partial\hat{u}_n}{\partial\theta}\right\|_{H^{3/2}_{\mathcal{D}}}$$

Also by Lemma 1

(11) $$\left\|\frac{\partial}{\partial r}\hat{\Omega}^\alpha(u_n)\right\|_{H^{3/2}_{\mathcal{D}}} \leq \text{const}\|u_n\|_{H^2}$$

(12) $$\left\|\frac{\partial}{\partial\theta}\hat{\Omega}^\alpha(u_n)\right\|_{H^{3/2}_{\mathcal{D}}} \leq \text{const}\|u_n\|_{H^2} .$$

Restricting all functions to S^1 inequalities (10), (11), and (12) become

(10') $$\left\|\frac{\partial\psi_n}{\partial r}\right\|_{H^1} \leq \text{const}\|u_n\|_{H^2} \cdot \left\|\frac{\partial u_n}{\partial\theta}\right\|_{H^1}$$

(11') $$\left\|\frac{\partial\Omega^\alpha}{\partial r}(u_n)\right\|_{H^1} \leq \text{const}\|u_n\|_{H^2}$$

(12') $$\left\|\frac{\partial\Omega^\alpha}{\partial\theta}(u_n)\right\|_{H^1} \leq \text{const}\|u_n\|_{H^2}$$

and clearly

(13) $$\left\|\frac{\partial u_n}{\partial\theta}\right\|_{H^1} \leq \|u_n\|_{H^2}$$

(14) $$\left\|\frac{\partial u_n}{\partial r}\right\|_{H^1} \leq \|u_n\|_{H^2} .$$

On S^1 equality (9) becomes

(9') $$\frac{\partial^2 u_n}{\partial r \partial \theta} = \frac{\partial \Omega^\alpha}{\partial r}(u_n)\left[\frac{\partial u_n}{\partial \theta}\right] - \frac{\partial \Omega^\alpha}{\partial \theta}(u_n)\left[\frac{\partial u_n}{\partial r}\right] + \frac{\partial}{\partial \theta}\left(\Omega^\alpha(u_n)\frac{\partial u_n}{\partial r}\right) - \frac{\partial \psi_n}{\partial r} .$$

Since $\{u_n\}$ is bounded in H^2, inequalities (10'), (11'), and (12') say that $\frac{\partial \psi_n}{\partial r}$, $\frac{\partial \Omega^\alpha}{\partial r}(u_n)$, and $\frac{\partial \Omega^\alpha}{\partial r}(u_n)$ are bounded in H^1. By the Sobolev embedding theorem the inclusion of $H^s(S^1, \mathbb{R}^m)$, $s > 1/2$, into the continuos functions $C(S^1, \mathbb{R}^m)$ is completely continuous. Therefore there is a subsequence of $\{u_n\}$ call it again $\{u_n\}$ so that $\frac{\partial \psi_n}{\partial r}$, $\frac{\partial \Omega^\alpha}{\partial r}(u_n)$, and $\frac{\partial \Omega^\alpha}{\partial r}(u_n)$, $\frac{\partial u_n}{\partial r}$, $\frac{\partial u_n}{\partial \theta}$, and u_n converge uniformly as continuous functions on S^1. From (1) we know that $\frac{\partial}{\partial \theta}(\Omega^\alpha(u_n)\frac{\partial u_n}{\partial r})$ converges to 0 in the H^0 (or L_2 norm). Equality (9') then says that $\frac{\partial^2 u_n}{\partial r \partial \theta}$ converges in H^0. This together with the fact that u_n, $\frac{\partial u_n}{\partial r}$ and $\frac{\partial u_n}{\partial \theta}$ converge in H^0 implies that u_n converges in $H^2(S^1, \mathbb{R}^3)$. This establishes Theorem 1.

COROLLARY. On any closed bounded subset $B \subset \mathcal{N}_\alpha$ the set of zeros of X^α in B is compact.

In §I we introduced the notion of a proper family of vertical vector fields. We would now like to introduce the notion of a b-proper (or bounded proper) family of vector fields.

Let $\pi : \mathcal{A} \times \mathcal{N}_0 \to \mathcal{A}$ be a product Banach fibre bundle over a Banach manifold $\mathcal{A}$. For each $n \in \mathcal{N} = \mathcal{A} \times \mathcal{N}_0$, $n = (a, n_0)$, $T_n\mathcal{N} = T_a\mathcal{A} \times T_{n_0}\mathcal{N}_0$. A vector field on $\mathcal{N}$ is vertical if for all (a, n_0), $X(a, n_0) \in T_{n_0}\mathcal{N}_0$. If

$\mathcal{N}_0$ has a Finsler structure $\| \ \| : T\mathcal{N}_0 \to R$ we have a metric on $\mathcal{N}_0$ and hence the notion of a bounded set.

DEFINITION. A vertical family on a product bundle $\mathcal{N}$ is b-proper if whenever $p_i = (a_i, q_i) \in \mathcal{N}$ is a sequence with $\|X(p_i)\| \to 0$, $\{q_i\}$ bounded, and $a_i \to a \in \mathcal{A}$, then $\{p_i\}$ has a convergent subsequence.

Using this local notion of b-proper we can now present a global notion.

DEFINITION. Let $\pi : \mathcal{N} \to \mathcal{A}$ be a smooth Banach fibre bundle with $\| \ \| : v(T\mathcal{N}) \to R$ a Finsler on the vertical subbundle of $T\mathcal{N}$. A vertical family $X : \mathcal{N} \to T\mathcal{N}$ is b-proper if for every $a_0 \in \mathcal{A}$ there is a neighborhood $\mathcal{U}$ of a_0 and a trivialization τ of $\mathcal{N}$ over $\mathcal{U}$, $\tau : \mathcal{U} \times \mathcal{N}_{a_0} \to \mathcal{N}$, $\mathcal{N}_{a_0} = \tau^{-1}(a_0)$, so that the vertical family X_τ on $\mathcal{U} \times \mathcal{N}_{a_0}$ given by $D\tau^{-1} X \circ \tau = X_\tau$ is b-proper on $\mathcal{U} \times \mathcal{A}_0$.

For the Plateau problem each $\mathcal{N}_\alpha$ is a submanifold of $H^2(S^1, \mathbb{R}^3)$ and so each $T_n \mathcal{N}_\alpha$ is a subspace of this Hilbert space. Consequently the H^2 norm induces a Finsler on the vertical subbundle $v(T\mathcal{N}) = \bigcup_{\substack{u \in \mathcal{N}_\alpha \\ \alpha \in a}} T_n \mathcal{N}_\alpha$ of $T\mathcal{N}$. A slight modification of the proof of Theorem 1 yields

THEOREM 2. If $\mathcal{N} = \bigcup \mathcal{N}_\alpha$ denotes the surface fibre bundle and X the minimal surface vector field on $\mathcal{N}$, then X is b-proper with respect th the $H^2(S^1, \mathbb{R}^3)$ Finsler induced on the vertical subbundle $v(T\mathcal{N})$ of $T\mathcal{N}$.

§XI. NON-DEGENERATE CRITICAL SUBMANIFOLDS OF $\mathcal{N}_\alpha$ AND A UNIQUENESS THEOREM FOR MINIMAL SURFACES

In this section we present a definition of non-degenerate minimal surface and show that any plane curve spans a unique non-degenerate minimal surface. We then use this fact to conclude that any curve sufficiently close to a plane curve has a unique minimal surface spanning it.

Only three other cases are known where uniqueness holds for the Plateau problem. If Γ is a plane curve the fact that there exists only one simply connected minimal surface spanning it follows from the Riemann-Osgood mapping theorem. If Γ has a simply covered convex curve as its parallel projection on a plane, then Rado [26] showed that there exists only one minimal surface spanning Γ. Recently [18] Nitsche showed that uniqueness holds if the total curvature of Γ does not exceed 4π.

Let $u \in \mathcal{N}_\alpha$ be a zero of X^α. Then the orbit $\mathcal{O}_G(u)$ of u under the induced action of the conformal group on $\mathcal{N}_\alpha$ consists of minimal surfaces (cf. §VII).

By Theorem 2, §VII, it follows that $\mathcal{O}_G(u)$ is the image of an immersion, and if $u \in \overline{\mathcal{M}}_\alpha$ this immersion is injective. According to the convention of some authors $\mathcal{O}_G(u)$ is then a submanifold of $\mathcal{N}_\alpha$, and using this terminology for the moment we see that simply connected minimal surfaces are never isolated but occur as families of three dimensional submanifolds of $\mathcal{N}_\alpha$. We shall call such manifolds <u>critical submanifolds</u> of $\mathcal{N}_\alpha$. Each critical point $u \in \overline{\mathcal{M}}_\alpha$ belongs to the critical submanifold $\mathcal{O}_G(u)$.

A minimal surface u spanning Γ is <u>unique</u> if the only other minimal surfaces spanning Γ are in $\mathcal{O}_G(u)$.

In §VII we showed that the tangent space $T_u[\mathcal{O}_G(u)]$ to the orbit of u under G at the point u is spanned by the three maps u_θ, $\sin\theta \cdot u_\theta$, and $\cos\theta \cdot u_\theta$. Since X^α is equivariant under the action of G it follows that $T_u[\mathcal{O}_G(u)] \subset \operatorname{Ker} X^\alpha_*(u)$, the kernel of $X^\alpha_*(u)$. Let $T_u[\mathcal{O}_G(u)]^\perp$ be the orthogonal complement of $T_u[\mathcal{O}_G(u)]$ in $T_u\mathcal{N}_\alpha$ with respect to the weak Riemannian structure $\langle\langle\, ,\, \rangle\rangle$. Using the fact that $DE_\alpha(u)(h) = \langle\langle X^\alpha(u), h\rangle\rangle$ one can easily show that $X^\alpha_*(u) : T_u\mathcal{N}_\alpha \circlearrowleft$ is self adjoint with respect to the weak inner product. Consequently $X^\alpha_*(u)$ maps $T_u[\mathcal{O}_G(u)]^\perp$ into itself.

DEFINITION. A minimal surface $u \in \mathcal{N}_\alpha$, or its orbit $\mathcal{O}_G(u)$, is <u>non-degenerate</u> if

$$X^\alpha_*(u) : T_u[\mathcal{O}_G(u)]^\perp \longrightarrow T_u[\mathcal{O}_G(u)]^\perp$$

is an isomorphism. If $u \in \mathcal{M}_\alpha$ we say that $\mathcal{O}_G(u)$ is a non-degenerate critical submanifold.

It is not hard to see that if $v = g_\#(u)$ is another point in $\mathcal{O}_G(u)$ then $X^\alpha_*(v) : T_v[\mathcal{O}_G(u)]^\perp \circlearrowleft$ is an isomorphism iff $X^\alpha_*(u) : T_u[\mathcal{O}_G(u)]^\perp \circlearrowleft$ is an isomorphism. Let $\mathcal{N}(\mathcal{O}_G(u)) \longrightarrow \mathcal{O}_G(u)$ denote the normal bundle of the orbit space $\mathcal{O}_G(u)$. Then X^α_* induces a bundle homomorphism of $\mathcal{N}(\mathcal{O}_G(u))$ into itself. Using this terminology we can formulate the following equivalent definition of non-degenerate minimal surface:

u is a non-degenerate minimal surface spanning $\Gamma = \alpha(S^1)$ if X^α_* induces a bundle isomorphism of $\mathcal{N}(\mathcal{O}_G(u))$ onto itself.

This notion of non-degeneracy depends only on the image curve Γ and not on any particular parameterization α for Γ (cf. remarks of §IX).

Let $\Gamma = S^1 \subset \mathbb{R}^2$ be the unit circle and $\alpha : S^1 \circlearrowleft$ the identity parameterization. There is only one minimal surface spanning the plane curve S^1 and we can take $u(z) = z$, z the complex coordinate of $\mathcal{D}$, to represent this surface (remember we are just picking one member of an entire orbit of minimal surfaces).

THEOREM 1. The minimal surface $u(z) = z$ is a non-degenerate minimal surface spanning $S^1 \subset \mathbb{R}^2$.

Proof. Write u in polar coordinates as $u(r, \theta) = (r \cos \theta, r \sin \theta)$. For $r = 1$, $u_\theta = (-\sin \theta, \cos \theta)$. Using the fact that $\sin \theta = \frac{e^{i\theta}-e^{-i\theta}}{2i}$, $\cos \theta = \frac{e^{i\theta}+e^{-i\theta}}{2}$ we see that $u_\theta = 1/2(1/i[-e^{i\theta}+e^{-i\theta}], [e^{i\theta}+e^{-i\theta}])$. Every $k \in T_u \mathcal{N}_\alpha$ can be represented by $k(\theta) = \lambda(\theta) \cdot u_\theta$, $\lambda : S^1 \longrightarrow \mathbb{R}$ an H^1 function. Let λ be represented by its Fourier series $\lambda(\theta) = \sum_\nu b_\nu e^{i\nu\theta}$ where $b_{-\nu} = \bar{b}_\nu$. Therefore

$$(1)\quad k(\theta) = \lambda(\theta)u_\theta = 1/2(-\frac{1}{i}\Sigma b_m e^{(m+1)\theta} + \frac{1}{i}\Sigma b_m e^{(m-1)\theta}, \Sigma b_m e^{(m+1)\theta} + \Sigma b_m e^{(m-1)\theta})$$

$$= 1/2(\frac{1}{i}\sum_\nu(b_{\nu+1}-b_{\nu-1})e^{i\nu\theta}, \sum_\nu(b_{\nu-1}+b_{\nu+1})e^{i\nu\theta}) .$$

Consequently

$$(2)\quad \frac{\partial}{\partial r}(\lambda(\theta)u_\theta) = 1/2(\frac{1}{i}\Sigma(b_{\nu+1}-b_{\nu-1})|\nu|e^{i\nu\theta}, \Sigma(b_{\nu-1}+b_{\nu+1})|\nu|e^{i\nu\theta} .$$

In this last equation we have used the fact that if $h(\theta) = \Sigma c_\nu e^{i\nu\theta}$, then its harmonic extension to $\mathcal{D}$ is given by $h(r, \theta) = \Sigma c_\nu r^{|\nu|} e^{i\nu\theta}$ and $\frac{\partial h}{\partial r} = \Sigma c_\nu |\nu| e^{i\nu\theta}$.

In order to show that $u(z) = z$ is non-degenerate it suffices to show that

(3) $$\langle\!\langle X_*(u)[k], k \rangle\!\rangle = 0$$

implies $k \in T_u[\mathcal{O}_G(u)]$; that is, $\lambda(\theta)$ is a linear combination of 1, $\sin\theta$, and $\cos\theta$. Therefore if $k(\theta) = \lambda(\theta)u_\theta$ satisfies (3) we must show that all the Fourier coefficients b_ν of λ vanish if $|\nu| > 1$. From §IX we know that

(4) $$\Omega^\alpha(u)\frac{\partial}{\partial r}X^\alpha_*(u)[k] = \Omega^\alpha(u)\frac{\partial k}{\partial r} - \Omega^\alpha(u)\frac{\partial}{\partial\theta}\left\{N(\theta)\langle k(\theta), U(\theta)\rangle\right\}$$

where $N(\theta) = \dfrac{u_r}{\|u_r\|}$, and $U(\theta) = \dfrac{u_\theta}{\|u_\theta\|}$. If $\langle\!\langle X^\alpha_*(u)[k], k \rangle\!\rangle = 0$ then

$$0 = \int_{S^1} \left\langle \frac{\partial}{\partial r}X^\alpha_*(u)[k], k\right\rangle d\theta = \int_{S^1} \left\langle \Omega^\alpha(u)\frac{\partial}{\partial r}X^\alpha_*(u)[k], k\right\rangle d\theta \ .$$

Plugging in (4) we find that

$$\begin{aligned}\int_{S^1} \left\langle \frac{\partial k}{\partial r}, k\right\rangle d\theta &= \int_{S^1} \left\langle \frac{\partial}{\partial\theta}\left\{N(\theta) \ \langle k(\theta), U(\theta)\rangle\right\}, k\right\rangle d\theta \\ &= \int_{S^1} \left\langle \frac{\partial}{\partial\theta}(\lambda(\theta)u_r), k(\theta)\right\rangle d\theta \\ &= \int_{S^1} \left\langle \frac{\partial}{\partial\theta}(\lambda(\theta)u_r), \lambda(\theta)u_\theta\right\rangle d\theta \\ &= \int_{S^1} |\lambda(\theta)|^2 \ \langle u_{r\theta}, u_\theta\rangle \ d\theta \ .\end{aligned}$$

In this last equality we used the fact that $\langle u_r, u_\theta\rangle = 0$. A short computation yields $\langle u_{r\theta}, u_\theta\rangle = 1$ and so the above term is equal to

(5) $$\int_{S^1}|\lambda(\theta)|^2 d\theta = 2\pi \sum_\nu |b_\nu|^2 \ .$$

But

$$\int_{S^1} \left\langle \frac{\partial k}{\partial r}, k\right\rangle d\theta = \int_{S^1} \left\langle \frac{\partial}{\partial r}(\lambda(\theta)u_\theta), \lambda(\theta)u_\theta\right\rangle d\theta$$

which by (1) and (2) is equal to

$$\frac{1}{4}\int_{S^1}\left\{\sum_{\nu,\omega}(b_{\nu+1}-b_{\nu-1})(b_{\omega+1}-b_{\omega-1})|\nu|e^{i(\nu+\omega)\theta}\right.$$

$$\left.+\sum_{\nu,\omega}(b_{\nu-1}+b_{\nu+1})(b_{\omega-1}+b_{\omega+1})|\nu|e^{i(\nu+\omega)\theta}\right\}d\theta$$

Using the relation $b_{-r} = \overline{b}_r$ this reduces to

(6)
$$\frac{\pi}{2}\left\{\sum_{\nu}|b_{\nu+1} - b_{\nu-1}|^2|\nu| + \sum_{\nu}|b_{\nu-1} + b_{\nu+1}|^2|\nu|\right\}$$
$$= \pi\left\{\sum_{\nu}(|b_{\nu+1}|^2 + |b_{\nu-1}|^2)|\nu|\right\}$$
$$= \pi\left\{\sum|b_\nu|^2(|\nu-1| + |\nu+1|)\right\} = 2\pi\sum_{\nu}|b_\nu|^2|\nu| .$$

Equating the results of (5) and (6) we must have

$$2\pi\sum_{\nu}|b_\nu|^2|\nu| = 2\pi\sum_{\nu}|b_\nu|^2 .$$

This implies that $b_\nu = 0$ for $|\nu| > 1$ and hence that $k \in T_u[\mathcal{O}_G(u)]$. Therefore the identity map of the disc represents a non-degenerate minimal surface spanning S^1, which completes the proof of Theorem 1.

We mentioned earlier that the Riemann-Osgood mapping theorem implies that any simple closed plane curve Γ has a unique minimal surface spanning it. If Γ is the unit circle this can be represented by the identity map of the disc into itself, and this map represents a non-degenerate minimum for the Dirichlet functional. Our next goal is to show that the unique minimal surface spanning a plane curve is non-degenerate.

A minimal surface spanning a plane curve is given by a complex analytic map $x : \mathcal{D} \longrightarrow \mathbb{C}$, $x(S^1) = \Gamma$. By Kellogg's theorem, if Γ is of Hölder class $C^{k,\rho}$, $k > 0$, then x is of class $C^{k,\rho}(\mathcal{D})$ with $x'(z) \neq 0$ for $|z| \leq 1$ (cf. Tsuji [31], page 361). Therefore x has an analytic inverse g mapping $x(\mathcal{D})$ back to $\mathcal{D}$.

Let $\mathcal{N}_\alpha$ be the manifold of harmonic surfaces spanning Γ with X^α the corresponding minimal surface vector field. By the equivariance of X^α under the action of G we know that the kernel of $X_*^\alpha(x)$ will be at least three dimensional. If we show that it is precisely three dimensional then we will have shown that x is non-degenerate.

Write $x(\theta) = (x^1(\theta), x^2(\theta))$. Since x is analytic the Cauchy Riemann equations state that on S^1, $\frac{\partial x^2}{\partial r} = \frac{\partial x^1}{\partial \theta}$, and $\frac{\partial x^1}{\partial r} = -\frac{\partial x^2}{\partial \theta}$.

Suppose that $X_*^\alpha(x)k = 0$ for some $k \in T_x\mathcal{N}_\alpha$. Let $k(\theta) = \lambda(\theta) \cdot x_\theta = (\lambda x^1, \lambda x^2) = (k^1, k^2)$. Using formula (4) with x replacing u we find that $\langle\!\langle X_*^\alpha(x)k, k\rangle\!\rangle = 0$ implies that

$$0 = \int_{S^1} \left\langle \Omega^\alpha(x)X_*^\alpha(x)k, k\right\rangle d\theta = \int_{S^1}\left\langle \frac{\partial k}{\partial r}, k\right\rangle d\theta - \int_{S^1}\left\langle \frac{\partial}{\partial\theta}(\lambda x_r), k\right\rangle d\theta .$$

But $x_r = (x_r^1, x_r^2) = (-x_\theta^2, x_\theta^1)$ and so $\lambda x_r = (-\lambda x_\theta^2, \lambda x_\theta^1) = (-k^2, k^1)$ and the above string of equalities become

$$0 = \int_{S^1}\left\langle (k_r^1, k_r^2), (k^1, k^2)\right\rangle d\theta - \int_{S^1}\left\langle (-k_\theta^2, k_\theta^1), (k^1, k^2)\right\rangle d\theta$$

$$= \int_{S^1}\frac{\partial k^1}{\partial r}\cdot k^1 d\theta + \int_{S^1}\frac{\partial k^2}{\partial r}\cdot k^2 + \int_{S^1}\frac{\partial k^2}{\partial\theta}\cdot k^1 - \int \frac{\partial k^1}{\partial\theta}k^2$$

$$= \int_{S^1}\frac{\partial k^1}{\partial r}\cdot k^1 d\theta + \int_{S^1}\frac{\partial k^2}{\partial r}\cdot k^2 d\theta - 2\int_{S^1}\frac{\partial k^1}{\partial\theta}\cdot k^2 d\theta .$$

Let $k^1(\theta) = \Sigma c_\nu^{i\nu\theta}$, $\bar{c}_\nu = c_{-\nu}$ and $k^2(\theta) = \Sigma d_\nu^{i\nu\theta}$, $\bar{d}_\nu = d_{-\nu}$. Using this Fourier representation this chain of equalities transforms to

$$(7) \qquad 0 = \langle\!\langle X_*^\alpha(x)k, k\rangle\!\rangle = 2\pi\sum_\nu |\nu|\{|c_\nu|^2 + |d_\nu|^2\} - 4\pi\Sigma i\nu c_\nu d_{-\nu}$$

$$= 2\pi\{\sum_\nu |\nu|\{|c_\nu|^2 + |d_\nu|^2\} - 2\sum_\nu i\nu c_\nu d_{-\nu}\} .$$

By the standard Schwarz inequality

(8) $$|\Sigma i\nu c_\nu d_{-\nu}| \le \sqrt{\Sigma|\nu||c_\nu|^2}\cdot\sqrt{\Sigma|\nu||d_\nu|^2}\ .$$

Let $p = \Sigma|\nu||c_\nu|^2$, $q = \Sigma|\nu||d_\nu|^2$. Then

(9) $$p+q \ge 2\sqrt{p}\cdot\sqrt{q}$$

with equality holding if and only if $p = q$. Inequality (9) together with equalities (7) and (8) imply that the Hessian of the Dirichlet functional

(10) $$D^2E_\alpha(x)(k, k) = \langle\!\langle X_*^\alpha(x)k, k\rangle\!\rangle \ge 0\ .$$

The positivity of the Hessian reflects the fact that the minimal surface spanning a plane curve is a minimum for E_α . We shall now study inequality (8) more carefully. Let

$$a_\nu = \begin{cases} i\sqrt{\nu}c_\nu, & \nu \ge 0 \\ -i\sqrt{|\nu|}c_\nu, & \nu \le 0 \end{cases}$$

and

$$b_\nu = \begin{cases} \sqrt{\nu}\ d_\nu, & \nu \ge 0 \\ \sqrt{|\nu|}\ d_\nu, & \nu \le 0 \end{cases}\ .$$

Then $\Sigma i\nu c_\nu d_{-\nu} = \{a_\nu\}\cdot\{b_\nu\}$, the dot product of sequences defined by $\{a_\nu\}\cdot\{b_\nu\} = \Sigma a_\nu\bar{b}_\nu$. Again applying the Cauchy-Schwarz inequality wè have

(11) $$|\{a_\nu\}\cdot\{b_\nu\}| \le |\{a_\nu\}|_{\ell_2}\cdot|\{b_\nu\}|_{\ell_2}$$

with equality holding if $\{b_\nu\}$ is a scalar multiple of $\{a_\nu\}$. In this inequality $|\,|_{\ell_2}$ denotes the sequential ℓ_2 norm of a sequence.

Considering equation (7) we see that $X_*^\alpha(x)k = 0$ implies that $p+q = 2|\{a_\nu\}\cdot\{b_\nu\}|$ and therefore applying (11) and (8)

$$p+q = 2\,|\{a_\nu\}\cdot\{b_\nu\}| \le 2\,|\{a_\nu\}|_{\ell_2}\cdot|\{b_\nu\}|_{\ell_2} = 2\sqrt{p}\cdot\sqrt{q}\ .$$

From inequality (9) we must have $p+q = 2\sqrt{p}\cdot\sqrt{q}$ and so $|\{a_\nu\}\cdot\{b_\nu\}| = |\{a_\nu\}|_{\ell_2}\,|\{b_\nu\}|_{\ell_2}$. By the Schwarz inequality (11) we can conclude that the sequence $\{b_\nu\}$ is a scalar multiple of $\{a_\nu\}$. It is easy to see that the only possible such scalar multiple is 1. Therefore for $\nu \ge 0$, $\sqrt{\nu}d_{-\nu} = i\sqrt{\nu}c_\nu$ and for $\nu \le 0$ $\sqrt{\nu}d_\nu = -i\sqrt{\nu}c_\nu$; or

$$\text{(12)}\qquad \begin{aligned} c_\nu &= -id_\nu,\ \nu > 0 \\ c_\nu &= id_\nu,\ \nu < 0\ . \end{aligned}$$

These equalities imply that $\frac{\partial k^1}{\partial\theta} = \frac{\partial k^2}{\partial r}$, or more importantly that the harmonic extension of $k : S^1 \longrightarrow \mathbb{R}^3$, $k(\theta) \in T_{x(\theta)}\Gamma$, to the disc $\mathcal{D}$, is a complex analytic map on $\mathcal{D}$. Thus pulling everything together we obtain

THEOREM 2. Let $\Gamma = \alpha(S^1)$ be a plane curve with $X_*^\alpha(x)k = 0$ for $k \in T_x\mathcal{N}_\alpha$. Then the harmonic extension of k to $\mathcal{D}$ is a complex analytic map.

Using this crucial result we can now prove

THEOREM 3. If $x : \mathcal{D} \longrightarrow \mathbb{R}^2$ represents the unique minimal surface spanning a plane curve $\Gamma = \alpha(S^1)$, then x is non-degenerate.

Proof. By Theorem 2 the kernel of $X_*^\alpha(x)$ consists of vector fields $k(\theta)$ tangent to Γ at $x(\theta)$ whose harmonic extensions $\hat{k}$ to $\mathcal{D}$ are complex analytic mappings. Recall that $x : \mathcal{D} \longrightarrow \mathbb{C}$ is complex analytic and has a complex analytic inverse $g : x(\mathcal{D}) \longrightarrow \mathcal{D}$. Let $k \in \operatorname{Ker} X_*^\alpha(x)$.

Then we may pull $\hat{k}$ back to $\mathcal{D}$ by g obtaining complex analytic vector field $\hat{k}_{\#}$ on $\mathcal{D}$ with the property that $k_{\#}(\theta) \in T_{e^{i\theta}}S^1$.

Therefore, since x and g are invertible, it follows that the dimension of the kernel of $X_*^{\alpha}(x)$ is precisely the dimension of the space of complex analytic vector fields on $\mathcal{D}$ tangent to the identity map of S^1 on the boundary. But each such vector field $\hat{k}_{\#}$ can be integrated to a one parameter family of complex analytic maps of $\mathcal{D}$ into itself. Therefore each such $\hat{k}_{\#}$ can be obtained as the derivative of a one parameter subgroup of the group G of conformal maps of the disc to itself. Since G is three dimensional it immediately follows that the space of all such $\hat{k}_{\#}$ is three dimensional. This proves Theorem 3.

As a direct consequence of the proofs to Theorems 2 and 3 we have

COROLLARY. Let $x : \mathcal{D} \longrightarrow \mathbb{R}^2$ be a minimal surface spanning a plane curve $\Gamma = \alpha(S^1)$. Then the Hessian $D^2E_{\alpha}(k, k) \geq 0$ and equals zero iff $k \in T_x[\mathcal{O}_G(x)]$.

We know that minimal surfaces are never isolated in the strict sense; for if u is minimal any $y \in \mathcal{O}_G(u)$ is also minimal. Nevertheless we can introduce the notion of isolated minimal surface as follows.

DEFINITION. A minimal surface $u \in \mathcal{N}_{\alpha}$ is <u>isolated</u> if there is an open neighborhood $\mathcal{W} \supset \mathcal{O}_G(u)$ such that if $y \in \mathcal{W}$ and y is minimal then $y \in \mathcal{O}_G(u)$.

In the next section we shall show that non-degenerate minimal surfaces are isolated. Recall, that in order to distinguish between critical points of E_{α} in $\mathcal{M}_{\alpha}$ and other critical points we have often referred to the former as classical minimal surfaces. It is rare to see the conformal group G mentioned in contemporary papers on minimal surfaces. The reason is that, from one point of view, the group G can be conveniently "factored out" of the problem as follows. Let Q_1, Q_2, Q_3 be three fixed points on S^1. For $\rho \in \mathcal{A}$, an embedding, let $\mathcal{S}_\rho$ be the codimension three submanifold of $\mathcal{N}_\rho$ consisting of those maps in $\mathcal{N}_\rho$ which send Q_i to $\rho(Q_i)$. Each orbit of G intersects $\mathcal{S}_\rho$ in one and only one point. Consequently if we restrict the Dirichlet functional E_ρ to $\mathcal{S}_\rho$ the action of G will have been factored out. The reason why we cannot stick entirely with this method of dealing with the G-action on $\mathcal{N}_\rho$ is simply that, for x a minimal surface spanning $\rho(S^1)$, the codimension three tangent space $T_x\mathcal{S}_\rho$ does not have a weak orthogonal complement as a subspace of $T_x\mathcal{N}_\rho$. This problem will become more apparent in the next two sections.

We conclude this section with a proof of

THEOREM 4. Let $\Gamma^\alpha = \alpha(S^1)$, $\alpha \in \mathcal{A}$, be a plane curve, and let $x(\alpha) \in \mathcal{S}_\alpha$ denote the unique (now also unique in the traditional sense) minimal surface spanning Γ^α. Then for any embedding $\rho \in \mathcal{A}$, "sufficiently close" to α, there is a unique minimal surface, $x(\rho) \in \mathcal{S}_\rho \cap \mathcal{M}_\rho$, spanning $\Gamma^\rho = \rho(S^1)$. The correspondence $\rho \longrightarrow x(\rho)$ is C^{r-1} smooth and moreover $x(\rho)$ is free of branch points.

PROOF. Let Π^ρ_ω be the H^2 orthogonal projection of $T_\omega\mathcal{N}_\rho$ onto $T_\omega\mathcal{S}_\rho$, and define the C^{r-1} vector field $Y^\rho : \mathcal{S}_\rho \longrightarrow T\mathcal{S}_\rho$ by

$$Y^\rho(\omega) = \Pi^\rho_\omega X^\rho(\omega) . \tag{13}$$

The zeros of X^ρ on $\mathcal{S}^\rho$ will clearly also be zeros of Y^ρ; however since Y^ρ is defined in terms of the H^2 orthogonal projection the zeros of Y^ρ need not be the zeros of X^ρ. Since $x(\alpha) \in \mathcal{S}_\alpha$ is a minimal surface $X^\alpha(x(\alpha)) = 0$ and so $Y^\alpha(x(\alpha)) = 0$; moreover $Y^\alpha_*(x(\alpha)) : T\mathcal{S}_\alpha \circlearrowleft$ is given by

$$(14) \qquad Y^\alpha_*(x(\alpha)) = \Pi^\alpha_{x(\alpha)} X^\alpha_*(x(\alpha)) \mid T_{x(\alpha)}\mathcal{S}_\alpha .$$

Using the fact that $\frac{\partial}{\partial\theta}x(\alpha) \neq 0$ for any θ (Kellogg's theorem) one can easily verify that

$$(15) \qquad T_{x(\alpha)}[\mathcal{O}_G(x(\alpha))] \cap T_{x(\alpha)}\mathcal{S}_\alpha = 0$$

and consequently $\mathcal{S}_\alpha$ intersects the manifold $\mathcal{O}_G(x(\alpha))$ transversally. Since $x(\alpha)$ is non-degenerate, (14) and (15) imply that $Y^\alpha_*(x(\alpha))$ is an isomorphism. Consequently, by the inverse function theorem, there is a neighborhood $\mathcal{W} \subset \mathcal{S}_\alpha \cap \mathcal{M}_\alpha$ of $x(\alpha)$ with the property that Y^α has no other zeros in $\mathcal{W}$ other than $x(\alpha)$. Thus X^α also has $x(\alpha)$ as its unique zero in $\mathcal{W}$.

Fix a δ, $0 < \delta < 1$. Any $\alpha \in \mathcal{A} \subset H^{r+2}(S^1, \mathbb{R}^3)$, $r \geq 5$ belongs to the Hölder space $C^{6,\delta}(S^1, \mathbb{R}^3)$. Let $\frac{1}{2}R_0$ denote the $C^{6,\delta}$ Hölder norm of the embedding α. We can find a neighborhood $\mathcal{V}_\alpha \subset \mathcal{A}$ of α with the property that any $\rho \in \mathcal{V}_\alpha$ has Hölder norm less than R_0. By Theorem 1, §VIII, for $\rho \in \mathcal{V}_\alpha$ any minimal surface in $\mathcal{S}_\rho \cap \overline{\mathcal{M}}_\rho$ spanning Γ^ρ will have a $C^{6,\delta}$ Hölder norm bounded by some fixed number, which depends only on R_0. Therefore, for $\rho \in \mathcal{V}_\alpha$, all minimal surfaces in $\mathcal{S}_\rho \cap \overline{\mathcal{M}}_\rho \subset H^2(S^1, \mathbb{R}^3)$ will have H^2 norms bounded by some constant R.

Let $B^\alpha_R \subset \mathcal{S}_\alpha \cap \overline{\mathcal{M}}_\alpha$ be the intersection of the ball of radius R in $H^2(S^1, \mathbb{R}^3)$ with $\mathcal{S}_\alpha \cap \overline{\mathcal{M}}_\alpha$. By Theorem 1, §X, the vector field X^α satisfies condition (CV) and so there exists a $\mu > 0$ with

(16) $$\inf_{u \varepsilon B_R^\alpha - \mathcal{W}} \| X^\alpha(u) \| \geq \mu$$

where the norm $\| \ \|$ is the H^2 norm.

Recall that $\pi : \mathcal{N} \to \mathcal{A}$, $\mathcal{N} = \bigcup_{\rho \varepsilon \mathcal{A}} \mathcal{N}_\rho$, has the structure of a smooth fibre bundle over $\mathcal{A}$. If the neighborhood $\mathcal{V}_\alpha$ of α is sufficiently small there is a trivialization of $\mathcal{N}$, $\tau : \mathcal{V}_\alpha \times \mathcal{N}_\alpha \to \pi^{-1}(\mathcal{V}_\alpha)$ given by $\tau(\rho, u) = \rho(\alpha^{-1}(u))$. Consider the union $\mathcal{S} = \bigcup_\rho \mathcal{S}_\rho$. τ induces a trivialization of $\bar{\pi} = \pi | \mathcal{S}$ and consequently $\mathcal{S}$ is a subbundle of $\mathcal{N}$. It is important to observe that τ carries the fibres $\rho \times \mathcal{S}_\alpha$ to $\mathcal{S}_\rho$ and also takes $\rho \times \overline{\mathcal{M}}_\alpha$ to $\overline{\mathcal{M}}_\rho$.

For $\rho \varepsilon \mathcal{V}_\alpha$ we can use the trivialization τ to transform the vector fields X^ρ and Y^ρ to a parameterized family (with ρ the parameter) of vector fields $\tilde{X}^\rho$ and $\tilde{Y}^\rho$ on the manifolds $\mathcal{N}_\alpha$ and $\mathcal{S}_\alpha$ respectively in such a way that

$$\tilde{X}^\alpha = X^\alpha \quad \text{and} \quad \tilde{Y}^\alpha = Y^\alpha \ .$$

It follows that $\tilde{Y}^\alpha_*(x(\alpha))$ is also an isomorphism where the Fréchet derivative is taken only along the fibre $\mathcal{S}_\alpha$. Applying the inverse function theorem we find that if $\mathcal{V}_\alpha$ and $\mathcal{W}$ are chosen small enough then for $\rho \varepsilon \mathcal{V}_\alpha$, $\tilde{Y}^\rho$ (and hence $\tilde{X}^\rho$) will have a unique zero $\tilde{x}(\rho) \varepsilon \mathcal{W}$, with $\rho \to \tilde{x}(\rho)$ a C^{r-1} map. Applying Theorem 2, §X, one can generalize inequality (16); namely that if $\mathcal{V}_\alpha$ is small enough, then for all

(17) $$\inf_{u \varepsilon B_R - \mathcal{W}} \|\tilde{X}^\rho(u)\| \geq \mu/2 \ .$$

This inequality, together with the remarks above, imply that $\tilde{X}^\rho$ has only $\tilde{x}(\rho) \varepsilon \mathcal{W}$ as a zero in $B_R^\alpha \subset \mathcal{S}_\alpha \cap \overline{\mathcal{M}}_\alpha$. From Theorem 1, §VIII, all zeros of $\tilde{X}^\rho$ in $\mathcal{S}_\alpha \cap \overline{\mathcal{M}}_\alpha$ must be in B_R^α (no zeros can suddenly appear far away. Therefore $\tilde{x}(\rho)$ is the unique zero of $\tilde{X}^{\rho}$ in $\mathcal{S}_{\alpha} \cap \overline{\mathcal{M}}_{\alpha}$. Setting $x(\rho) = \tau[\tilde{x}(\rho)]$ we see that $x(\rho)$ is the unique zero of X^{ρ} in $\mathcal{S}_{\rho} \cap \overline{\mathcal{M}}_{\rho}$. Since $\tilde{x}(\rho) \in \mathcal{W} \subset \mathcal{M}_{\alpha}$, $x(\rho) \in \mathcal{M}_{\rho}$ which means that $x(\rho)$ is free of branch points. It is obvious from the proof that $\rho \longrightarrow x(\rho)$ is C^{r-1}.

REMARK. It would be nice to have a sharp analytical estimate in terms of the geometrical data of the plane curve Γ^{α}, on exactly "how close" ρ needs to be to α for uniqueness to hold.

§XII. THE SPRAY OF THE WEAK METRIC

In this section we prove the existence of a smooth spray and exponential map for the weak Riemannian structure on $T\mathcal{N}_\alpha$. These are used to construct local codimension three submanifolds transverse to the orbit of a zero of X^α. Such local submanifolds (or slices) are used in showing that non-degenerate minimal surfaces are isolated, and are also crucial to several theorems in the next section.

To begin with, let us briefly review the notion of a spray and exponential map on a Banach manifold. For more details see [15 IV. 3].

DEFINITION. A spray on a manifold M is a vector field on the tangent bundle $\pi : TM \longrightarrow M$ which satisfies

(i) $\pi_* \xi(v) = v$

(ii) $\xi(sv) = s_* s\xi(v)$

where $\pi_* : T^2(TM) \longrightarrow TM$ is the Fréchet derivative of the projection map π and s_* denotes the Fréchet derivative of the bundle isomorphism from TM to itself which is fibrewise multiplication by the real number s.

The following theorem characterizes how sprays look locally; i.e. in some coordinate chart.

THEOREM 1. Let $\mathcal{U}$ be open in the Banach space E viewed as a C^r manifold, and let $T(\mathcal{U}) = \mathcal{U} \times E$ be the tangent bundle. A C^{r-2} map $f : \mathcal{U} \times E \longrightarrow \mathcal{U} \times E \times (E \times E)$ is a local representation of a spray on $T(\mathcal{U})$ iff f is of the form

$$f(x, v) = (x, v, f_2(x, v))$$

where $f_2(x, sv) = s^2 f_2(x, v)$ for all real numbers s.

If M is a C^r Riemannian manifold the Riemannian structure induces a C^{r-2} spray on M, known as the geodesic spray. Generally this is constructed as follows:

On any C^r Hilbert manifold M there is a canonical and fundamental closed non-degenerate two-form ω on TM. In particular, ω is a C^{r-1} section of the tensor bundle over TM whose fibre over $p \in TM$ consists of the real valued, antisymmetric, bilinear real valued maps on $T_p(TM)$. Therefore for each $p \in TM$ we have an antisymmetric bilinear map $\omega_p : T_p(TM) \times T_p(TM) \longrightarrow R$. Let $\mathcal{J} : TM \longrightarrow R$ be defined by

$$\mathcal{J}(v) = \frac{1}{2} \langle v, v \rangle_x ,$$

where $\langle , \rangle_x : T_xM \times T_xM \longrightarrow R$ is the Riemannian structure on TM. Then $-d\mathcal{J}$ is a section of $T^*(TM)$. Let $n_p \in T_p(TM)$. Since the two form ω is non-degenerate the map $n_p \longrightarrow \omega_p(n_p, \cdot)$ induces a bijective correspondence between the one forms and vector fields on TM. The geodesic spray is then defined to be the vector field on TM corresponding to the one-form $-d\mathcal{J}$.

Suppose M is a Riemannian manifold, but also has an inner product $\langle\langle , \rangle\rangle : TM \times TM \longrightarrow R$ which is weaker than the Riemannian inner product. Moreover suppose one needs to construct a spray corresponding to the weak metric $\langle\langle , \rangle\rangle$. Then the above construction breaks down.

However, the appearance of weak metrics and their associated geodesic sprays (when they exist) in fundamental problems in non-linear analysis is not uncommon; in fact a detailed study of the spray of a weak metric is the key to the now famous solutions to the Euler and Navier-Stokes equations of hydrodynamics by Ebin and Marsden [6]. It would therefore seem desirable to have a theory of weak metrics and their associated geodesic sprays and such a theory is somewhat developed in [36].

We shall now give another, more intuitive construction of the geodesic spray in the special case that our Hilbert manifold M is isometrically embedded in a Hilbert space H. This alternate construction will enable us to see how to construct a spray on $\mathcal{H}_\alpha$ associated to the weak Riemannian structure introduced in §VI.

So, let us suppose M is a C^r closed Riemannian submanifold of a Hilbert space H, with M given the Riemannian structure induced by the Riemannian structure on H. For each $m \in M$ we get a map $\Pi_m : H \longrightarrow T_m M$ which is the orthogonal projection of H onto $T_m M$. The map $m \longrightarrow \Pi_m$ is a C^{r-1} map of M into $\mathcal{L}(H)$. The tangent bundle $T(H)$ and second tangent bundle $T^2(H)$ over H and $T(H)$ respectively are canonically isomorphic to $H \times H$ and $H \times H \times (H \times H)$. Define the map $Z : T(H) \longrightarrow T^2(H)$ by

$$Z(x, v) = ((x, v), (v, 0)) .$$

Let $\dot{j} : M \longrightarrow H$ be the inclusion. Then TM is a subbundle of the pull back bundle $\dot{j}^*[T(H)]$. Z can also be viewed as a map from $\dot{j}^*[T(H)]$ to $T(\dot{j}^* T(H))$ since these bundles are equivalent to $M \times H$ and $(M \times H) \times (H \times H)$ respectively.

Define the bundle map $\Pi_\# : \dot{j}^*[T(H)] \longrightarrow TM$ by

$$\Pi_\#(m, v) = (m, \Pi_m v) . \tag{1}$$

The map $\tilde{\Pi} : \dot{j}^*[T(H)] \longrightarrow H$, given by $\tilde{\Pi}(m, v) = \Pi_m v$, is the "principal part" of $\Pi_\#$.

Let $D\Pi_\# : T(\dot{j}^*[T(H)]) \longrightarrow T^2 M$ denote the tangent map of $\Pi_\#$. Now

$(D\Pi_{\#}) \circ Z : \mathcal{J}^*[T(H)] \longrightarrow T^2M$ restricts to a map from TM to T^2M. Clearly

$$D\Pi_{\#}(m, v, h, k) = (m, v, h, D\Pi_m(h)(v) + \Pi_m k)$$

where $D\Pi_m$ denotes the Fréchet derivative of $m \longrightarrow \Pi_m$. Therefore

$$(D\Pi_{\#} \circ Z)(m, v) = (m, v, v, D\Pi_m(v)(v)) .$$

One now easily checks that $\xi : TM \longrightarrow T^2M$ defined by

$$\xi(m, v) = (D\Pi_{\#} \circ Z)(m, v) \tag{2}$$

is a spray on M, and in fact is the geodesic spray induced by the Riemannian structure M inherits from H.

Using this construction we are now prepared to show that the weak Riemannian structure $\langle\langle , \rangle\rangle : T\mathcal{N}_\alpha \times T\mathcal{N}_\alpha \longrightarrow R$ on the C^r manifold $\mathcal{N}_\alpha$ induces a C^{r-2} spray on $T\mathcal{N}_\alpha$. At the end of §VI we showed that for each $u \in \mathcal{N}_\alpha$ there is a weak orthogonal projection $\Pi_u : H^2(S^1, \mathbb{R}^3) \longrightarrow T_u\mathcal{N}_\alpha$. Let $L_u : T_u\mathcal{N}_\alpha \longrightarrow (T_u\mathcal{N}_\alpha)^1$ be the isomorphism introduced in §V, where $L_u h = \Omega(u)\frac{\partial h}{\partial r}$ for $h \in T_u\mathcal{N}_\alpha$. Then $\Pi_u(k)$ for $k \in H^2(S^1, \mathbb{R}^3)$ was defined by

$$\Pi_n k = L_u^{-1}(\Omega(u)\frac{\partial k}{\partial r}) .$$

Since $u \longrightarrow L_u$, and $u \longrightarrow \Omega(u)$ is C^{r-1}, it follows that $u \longrightarrow \Pi_u$ is C^{r-1}.

Let $H = H^2(S^1, \mathbb{R}^3)$ and $M = \mathcal{N}_\alpha \subset H$. <u>Then formula</u> (2) <u>defines the geodesic spray</u> ξ <u>associated to the weak Riemannian structure</u>. As a vector field on TM, ξ is a C^{r-2} map and so even though $\langle\langle , \rangle\rangle$ is weaker than the H^2 inner product on $T\mathcal{N}_\alpha$ we still obtain a C^{r-2} spray.

To each spray $\xi : TM \longrightarrow T^2M$ on a manifold M there is an exponential map $\exp : \mathcal{U} \longrightarrow M$ which is defined on an open neighborhood $\mathcal{U}$ of the zero section of TM and which is constructed as follows. Let $t \longrightarrow \beta_v(t)$ be the flow on TM of the vector field ξ with initial condition $v \in TM$. Let $\mathcal{U}$ be the set of vectors on TM such that β_v is defined at least on the unit interval $[0, 1]$. It is standard [15] that $\mathcal{U}$ is an open neighborhood of the zero section of TM. If $\pi : TM \longrightarrow M$ denotes the tangent bundle projection map we define the map $\exp : \mathcal{U} \longrightarrow M$ by $\exp(v) = \pi\beta_v(1)$. If M is a C^r manifold $\exp$ is a C^{r-2} map. For $m \in M$ define $\exp_m = \exp \mid T_mM$. The map $\exp_m$ is a local diffeomorphism from a neighborhood of the zero $0_m \in T_mM$ to a neighborhood of m with $\exp_m(0_m) = m$ and $(\exp_m)_*(0_m)$, the Fréchet derivative of $\exp_m$ at 0_m, the identity map of T_mM into itself.

If $\sigma(t) = \pi\beta_v(t)$ one can show from the definition of a spray that σ satisfies the differential equation

$$\sigma''(t) = D\Pi_\sigma(\sigma')(\sigma') , \tag{3}$$

where $D\Pi_m$ again denotes the Fréchet derivative of $m \longrightarrow \Pi_m$. Moreover for such a σ, $\exp_m v = \sigma(1)$, where $\sigma(0) = m$, $\sigma'(0) = v$.

Let us formally state the main parts of the preceding results as

THEOREM 2. The weak Riemannian structure on $\mathcal{N}_\alpha$ induces a smooth C^{r-2} geodesic spray. The corresponding exponential map restricted to $T_u\mathcal{N}_\alpha$ gives a local diffeomorphism of a neighborhood of zero in $T_u\mathcal{N}_\alpha$ onto a neighborhood of u in $\mathcal{N}_\alpha$.

We now come to the principal reason for using the geodesic spray of the weak metric; namely the construction of a local C^{r-2} codimension three submanifold of $\mathcal{N}_\alpha$ transverse to the orbit of a minimal surface u. Let $u \in \overline{\mathcal{M}}_\alpha$ be a zero of X^α. By Theorem 2, §VII, the orbit $\mathcal{O}_G(u)$ of u under G is the image of an injective immersion. The tangent space $\mathbb{F}_u = T_u[\mathcal{O}_G(u)]$ is the three dimensional subspace of $T_u\mathcal{N}_\alpha$ spanned by u_θ, $\cos\theta \cdot u_\theta$, and $\sin\theta \cdot u_\theta$. Let $\mathbb{F}_u^\perp$ denote the orthogonal complement of $\mathbb{F}_u$ with respect to the weak inner product on $T_u\mathcal{N}_\alpha$. $\mathbb{F}_u^\perp$ is a codimension three subspace and $\exp_u$ maps a neighborhood of $0_u \in \mathbb{F}_u$ diffeomorphically onto a codimension three submanifold $\mathcal{E}$ of $\mathcal{N}_\alpha$ with $u \in \mathcal{E}^o$ and $\mathcal{E}$ transverse to $\mathcal{O}_G(u)$ at u. From the fact that for fixed $h \in H^2(S^1, \mathbb{R}^3)$ the map $v \longrightarrow D\Pi_m(v)(h)$ extends to a continous linear map in the $H^{1/2}$ topologies it follows that for any $w \in \mathcal{E}$, $T_w\mathcal{N} = T_w\mathcal{E} \oplus (T_w\mathcal{E})^\perp$, $(T_w\mathcal{E})$ the weak orthogonal complement of $T_w\mathcal{E}$.

Formally we have

THEOREM 3. If $u \in \mathcal{N}_\alpha$ is a zero of X^α then there is a local codimension three submanifold $\mathcal{E}$ of $\mathcal{N}_\alpha$ containing u as an interior point with the properties that $T_u\mathcal{E} = (T_u[\mathcal{O}_G(u)])^\perp$ and for $w \in \mathcal{E}$, $T_w\mathcal{N}_\alpha = T_w\mathcal{E} \oplus (T_w\mathcal{E})^\perp$.

$\mathcal{E}$ is called the <u>local exponential surface about a zero</u> u <u>of</u> X^α. Such an exponential surface provides a useful way of locally "factoring out" the action of G.

Recall from §VII that if $g \in G$, $g_\# : \mathcal{N}_\alpha \circlearrowleft$ denoted the induced diffeomorphism on $\mathcal{N}_\alpha$, and $g \longrightarrow A_g$ the induced H^2 representation of G. For a fixed g, A_g restricts to an isomorphism from $T_u\mathcal{N}_\alpha$ to $T_{g_\#(u)}\mathcal{N}_\alpha$.

The next theorem follows from the equivariance of the weak Riemannian structure under the action of G.

THEOREM 4. The restricted exponential map $\exp_u : T_u\mathcal{N}_\alpha \cap \mathcal{U} \longrightarrow \mathcal{N}_\alpha$ is <u>equivariant</u> in the following sense:

If $h \in \mathcal{U} \cap T_u\mathcal{N}_\alpha$, $\mathcal{U}$ the domain of exp , then $A_g h \in \mathcal{U} \cap T_{g_\#(u)}\mathcal{N}_\alpha$ and $g_\#(\exp_u h) = \exp_{g_\#(u)} A_g h$. Moreover if $h \in (T_u[\mathcal{O}_G(u)])^\perp$ then

$$A_g h \in (T_{g_\#(u)}[\mathcal{O}_G(u)])^\perp = (T_{g_\#(u)}[\mathcal{O}_G(g_\#(u))])^\perp .$$

As a consequence we obtain

THEOREM 5. If $\mathcal{E}$ is a local exponential surface about a zero u of X^α then for fixed $g \in G$, $g_\#(\mathcal{E}) = \{g_\#(w) \mid w \in \mathcal{E}\}$ is a local exponential surface about $g_\#(u)$.

THEOREM 6. Let N_{g_0} be a neighborhood of $g_0 \in G$ where N is a neighborhood of the identity of G and N_{g_0} is right translation of elements of N by g_0 . For a subset $A \subset \mathcal{N}_\alpha$ set $\mathcal{O}_{N_{g_0}}(A) = \{g_\#(u) \mid g \in N_{g_0} , u \in A\}$. If $u \in \overline{\mathcal{M}}_\alpha$ is a zero of X^α, then for a sufficiently small local exponential surface $\mathcal{E}$ about u , and a sufficiently small neighborhood N , $\mathcal{O}_{N_{g_0}}(\mathcal{E})$ is an open neighborhood of $(g_0)_\# u$; and moreover $\mathcal{O}_{N_{g_0}}(\mathcal{E})$ is diffeomorphic to the product $N_{g_0} \times \mathcal{E}$.

REMARK. Theorem 6 is actually quite surprising in view of the fact that for $\omega \in \mathcal{E}$, $\mathcal{O}_N(\omega)$ need not be a submanifold of $\mathcal{N}_\alpha$.

<u>Proof</u>. By the equivariance of the exponential map (Theorem 4) it suffices to consider only the case when g_0 is the identity of G . For a neighborhood N , and an exponential surface $\mathcal{E}$ about u , define a map

from $N \times \mathcal{E}$ to $\mathcal{N}_\alpha$ by

$$(g, \exp_u h) \longrightarrow \exp_{g_\#(u)} A_g h$$

where $h \in (T_u[\mathcal{O}_G(u)])^\perp$. To show that this map is a local diffeomorphism it is enough to show that the map from $N \times (T_u[\mathcal{O}_G(u)])^\perp$ to $\mathcal{N}_\alpha$ given by

$$(4) \qquad (g, h) \longrightarrow \exp_{g_\#(u)} A_g h$$

is a local diffeomorphism of a neighborhood of u.

By the inverse function theorem we need only demonstrate that the derivative of the map (4) at (id, 0), id the identity of G, is an isomorphism. But the partial derivative of (4) with respect to the h "variable" evaluated at (id, 0) is just the identity isomorphism of $(T_u[\mathcal{O}_G(u)])^\perp$ to itself (cf. earlier remarks on the derivative of the exponential map).

For $h = 0$, $\exp_{g_\#(u)} A_g h = g_\#(u)$. Therefore the partial derivative of (4) in the "G" direction sends three basis elements of $T_{id}G$ to u_θ, $\cos\theta \cdot u_\theta$, and $\sin\theta \cdot u_\theta$ which form a basis of $T_u[\mathcal{O}_G(u)]$ for details see §VII).

Combining these two partial derivatives we see immediately that the Fréchet derivative of (4) at (id, 0) is an isomorphism from $T_{id}G \times (T_u[\mathcal{O}_G(u)])^\perp$ to $T_u[\mathcal{O}_G(u)] \times (T_u[\mathcal{O}_G(u)])^\perp = T_u\mathcal{N}_\alpha$. This concludes the proof of Theorem 6.

Recall that a minimal surface u is isolated if there exists an open neighborhood $\mathcal{W}$ of $\mathcal{O}_G(u)$ in $\mathcal{N}_\alpha$ such that if $\omega \in \mathcal{W}$ and $X^\alpha(\omega) = 0$ then $\omega \in \mathcal{O}_G(u)$.

THEOREM 7. Let u be a zero of X^α in $\overline{\mathcal{M}}_\alpha$ and suppose that for some sufficiently small local exponential surface $\mathcal{E}$ through u there is no other zero of X^α in $\mathcal{E}$. Then u is isolated in the above sense.

PROOF. Since $u \in \overline{\mathcal{M}}_\alpha$, $g \longmapsto g_\#(u)$ is an injective immersion. By the preceding theorem, for fixed $g_0 \in G$, $\mathcal{O}_{N_{g_0}}(\mathcal{E}')$ is an open neighborhood of $(g_0)_\#(u)$ if N and $\mathcal{E}' \subset \mathcal{E}$ are sufficiently small where $\mathcal{E}'$ is an exponential surface about u.

Suppose $v \in \mathcal{O}_{N_{g_0}}(\mathcal{E}')$ is a zero of X^α. Then $v = (g^{-1}g_0)_\#(\omega)$, $\omega \in \mathcal{E}'$. Since X^α is equivariant under G, $\omega = (g_0^{-1}g)_\#(v)$ is also a zero of X^α in $\mathcal{E}$. By assumption $\omega = u$, and so $v \in \mathcal{O}_G(u)$. Setting $\mathcal{W}$ equal to the union of all such neighborhoods $\mathcal{O}_{N_{g_0}}(\mathcal{E}')$ gives us the required neighborhood of the orbit $\mathcal{O}_G(u)$.

We would now like to show that if u is a zero of X^α and $\mathcal{E}$ is a sufficiently small exponential surface containing u then there is a vector field Y^α on $\mathcal{E}$ with the property that the zeros of Y^α on $\mathcal{E}$ are precisely the zeros of X^α on $\mathcal{E}$.

If $\omega \in \mathcal{N}_\alpha$ then formally the tangent space $T_\omega[\mathcal{O}_G(\omega)] \subset (T_\omega\mathcal{N}_\alpha)^1$ and is spanned by the vectors ω_θ, $\cos\theta\cdot\omega_\theta$, and $\sin\theta\cdot\omega_\theta$; of course if ω is a zero of X^α then $T_\omega[\mathcal{O}_G(\omega)] \subset T_\omega\mathcal{N}_\alpha$.

Let $\mathcal{E}$ be a local exponential surface about a zero u of X^α. Then for $\omega \in \mathcal{E}$,

$$T_\omega\mathcal{N}_\alpha = T_\omega\mathcal{E} \oplus (T_\omega\mathcal{E})^\perp .$$

Let P_ω be the weak orthogonal projection of $T_\omega\mathcal{N}_\alpha$ onto $T_\omega\mathcal{E}$ with $\tilde{P}_\omega = I - P_\omega$, I the identity linear map on $T_\omega\mathcal{N}_\alpha$.

The vector field Y^α on $\mathcal{E}$ defined by $Y^\alpha(\omega) = P_\omega X^\alpha(\omega)$ is a C^{r-3} vector field.

THEOREM 8. If $\mathcal{E}$ is a sufficiently small exponential surface about a zero u of X^α then the zeros of Y^α are precisely the zeros of X^α on $\mathcal{E}$.

PROOF. For $\omega \in \mathcal{E}$, let $(T_\omega \mathcal{E})^1$ denote the closure of $T_\omega \mathcal{E}$ in $(T_\omega \mathcal{N}_\alpha)^1$. Since $\langle\langle\, ,\, \rangle\rangle$ is continuous in the H^1 topology we see that

$$T_u \mathcal{N}_\alpha = T_u \mathcal{E} \oplus T_u[\mathcal{O}_G(u)]$$

and

$$(T_u \mathcal{N}_\alpha)^1 = (T_u \mathcal{E})^1 \oplus T_u[\mathcal{O}_G(u)]$$

where both $T_u \mathcal{E}$ and $(T_u \mathcal{E})^1$ are weak perpendicular to $T_u[\mathcal{O}_G(u)]$. If $\omega \in \mathcal{E}$ is sufficiently close to u then

$$(5) \qquad (T_\omega \mathcal{N}_\alpha)^1 = (T_\omega \mathcal{E})^1 \oplus T_\omega[\mathcal{O}_G(\omega)]$$

and so

$$(6) \qquad (T_\omega \mathcal{E})^1 \cap T_\omega[\mathcal{O}_G(\omega)] = 0$$

LEMMA. For any $\omega \in \mathcal{N}_\alpha$ the vector field X^α is perpendicular to $T_\omega[\mathcal{O}_G(\omega)]$ with respect to the weak inner product.

<u>Proof</u> (of lemma). Recall that the Dirichlet functional $E_\alpha : \mathcal{N}_\alpha \longrightarrow R$ has Fréchet derivative

$$DE_\alpha(\omega)[h] = \langle\langle X^\alpha(\omega), h \rangle\rangle.$$

Let g_t be a C^∞ path in G with g_0 the identity. Then for $\omega \in \mathcal{N}_\alpha$, $t \longrightarrow (g_t)_\#(\omega)$ is a C^1 map of R into $H^1(S^1, \mathbb{R}^3)$ with

$$\frac{d}{dt}[(g_t)_\#(\omega)]_{t=0} = h \in T_\omega[\mathcal{O}_G(\omega)] .$$

By the equivariance of E under G

$$E(\omega) = E((g_t)_\#(\omega)) .$$

Therefore

$$0 = \frac{d}{dt} E((g_t)_\#(\omega)) = DE((g_t)_\#(\omega))[\frac{d}{dt}[(g_t)_\#(\omega)]] .$$

Evaluating at $t = 0$ we get that $\langle\langle X^\alpha(\omega), h\rangle\rangle = 0$. Since any $h \in T_\omega[\mathcal{O}_G(\omega)]$ can be obtained in this manner the proof of the lemma is complete, and we return to the proof of Theorem 8.

For $\omega \in \mathcal{E}$,

$$X^\alpha(\omega) = P_\omega X^\alpha(\omega) + \tilde{P}_\omega X^\alpha(\omega) \tag{7}$$

$$= Y^\alpha(\omega) + \tilde{P}_\omega X^\alpha(\omega) .$$

Let $\mathcal{E}$ be small enough so that for $\omega \in \mathcal{E}$ conditons (5) and (6) hold. Then

$$X^\alpha(\omega) = X_1^\alpha(\omega) + X_2^\alpha(\omega) , \tag{8}$$

where $X_1^\alpha(\omega) \in (T_\omega \mathcal{E})^1$, and $X_2^\alpha(\omega) \in T_\omega[\mathcal{O}_G(\omega)]$. Thus equating (7) and (8)

$$X_1^\alpha(\omega) + X_2^\alpha(\omega) = Y^\alpha(\omega) + \tilde{P}_\omega X^\alpha(\omega) .$$

Now suppose that $Y^\alpha(\omega) = 0$. Then

$$\langle\!\langle \tilde{P}_\omega X^\alpha(\omega), \tilde{P}_\omega X^\alpha(\omega) \rangle\!\rangle = \langle\!\langle X_1^\alpha(\omega) + X_2^\alpha(\omega), \tilde{P}_\omega X^\alpha(\omega) \rangle\!\rangle$$

$$= \langle\!\langle X_1^\alpha(\omega), \tilde{P}_\omega X^\alpha(\omega) \rangle\!\rangle + \langle\!\langle X_2^\alpha(\omega), X^\alpha(\omega) \rangle\!\rangle ,$$

since $X^\alpha(\omega) = \tilde{P}_\omega X^\alpha(\omega)$. By the lemma $\langle\!\langle X_2^\alpha(\omega), X^\alpha(\omega) \rangle\!\rangle = 0$. Since $\tilde{P}_\omega$ is the projection onto the weak orthogonal complement of $T_\omega \mathcal{E}$ we see that $\langle\!\langle X_1^\alpha(\omega), \tilde{P}_\omega X^\alpha(\omega) \rangle\!\rangle = 0$. Therefore $\tilde{P}_\omega X^\alpha(\omega) = 0$ which by (7) implies that $X^\alpha(\omega) = 0$.

If $X^\alpha(\omega) = 0$ it is trivial that $Y^\alpha(\omega) = 0$. Thus X^α and Y^α have precisely the same zeros.

We end this section with

THEOREM 9. Non-degenerate minimal surfaces are isolated.

Proof. Let $u \in \overline{\mathcal{M}}_\alpha$ be non-degenerate. We must show that $\mathcal{O}_G(u)$ is contained in an open set $\mathcal{W}$ which contains no other zeros of X^α other than those in $\mathcal{O}_G(u)$. By Theorem 7 it suffices to show that a sufficiently small exponential surface $\mathcal{E}$ about u contains no other zeros of X^α .

Let $\mathcal{E}'$ be small enough so that the vector field Y^α constructed in Theorem 8 has precisely the same zeros as X^α . Since u is non-degenerate it follows that

$$P_u X_*^\alpha(u) \mid T_u \mathcal{E}' = Y_*^\alpha(u) : T_u \mathcal{E}' \circlearrowleft$$

is an isomorphism. Therefore by the inverse function theorem u must be an isolated zero of Y^α (and hence of X^α) in $\mathcal{E}'$. Consequently if we take an exponential surface $\mathcal{E} \subset \mathcal{E}'$ small enough u would be the only zero of X^α on $\mathcal{E}$.

§XIII. THE TRANSVERSALITY THEOREM

The main goal of this section is to prove that there is an open set $\mathcal{F}$ of embeddings of S^1 into $\mathbb{R}^3$ (the fine embeddings) and an open dense set $\mathcal{V}$ of $\mathcal{F}$ with the property that if $\alpha \in \mathcal{V}$ then the number of classical simply connected minimal surfaces spanning the image Γ^α of α is finite in number.

If $\alpha \in \mathcal{V}$ and γ is sufficiently close to α, then the number of minimal surfaces spanning Γ^γ is equal to the number spanning Γ^α. This generalizes Theorem 4 of §XI.

In §IX we introduced the surface fibre bundle $\mathcal{N}$ over the space of embeddings $\mathcal{A}$ and later used this notion in §XI to prove a concrete analytical result. In §I we introduced the notion of a vertical family of vector fields and the concept of such a family being transverse to the zero section of the vertical subbundle of $T\mathcal{N}$. Our goal is to use Theorem 2 §I to prove the open density result; but we must proceed carefully. The vertical family X^α of minimal surface vector fields cannot, however, be transverse to the zero section of the vertical subbundle of $T\mathcal{N}$ because of the action of the conformal group. To see this, note that if $u \in \mathcal{N}_\alpha$ is a zero of X^α then the total Fréchet derivative of the vector field X at u, $X_*(u)$, is a linear map into $T_u\mathcal{N}_\alpha$ and its range is always contained in the weak orthogonal complement of $T_u[\mathcal{O}_G(u)] \subset T_u\mathcal{N}_\alpha$.

Consequently in order to apply a result similar to Theorem 2 of §I we must use some method to factor out the action of G. We begin by establishing some terminology.

Again let $w = x + iy$ (cf. §VIII) denote the complex coordinate system for $\mathbb{R}^2$. Recall that if $u \in \mathcal{N}_\alpha$ is a minimal surface then the component functions of u appear as the real parts of functions f_j, analytic on $\mathcal{D}^0$, $C^{5,\delta}$ on $\mathcal{D}$, $f_j(w) = u_j(w) + i\,\tilde{u}_j(w)$, $j = 1,2,3$ where $\tilde{u}_j$ represents the harmonic conjugate of u_j and the f_j satisfy Monge's relation $\sum_j [f_j'(w)]^2 = 0$.

A point $p \in \mathcal{D}$ is a branch point of u if $f_j'(p) = 0$ for $j = 1,2,3$. By the <u>order</u> of a branch point p we mean the minimum over all j of the order of p as a zero of $f_j'(p)$. By the corollary to Theorem 2 §VII we know that each minimal surface $u \in \overline{\mathcal{M}}_\alpha$ can have at most a finite number of branch points.

For a fixed $\alpha \in \mathcal{A}$ we know that the vector field X^α is Palais-Smale (c.f. §V) and consequently the codimension of $X_*^\alpha(u)[T_u\mathcal{N}_\alpha]$ in $T_u\mathcal{N}_\alpha$ at a zero u is finite. In general, we have no control over the magnitude of this dimension. However, if we look at the image of the total Fréchet derivative $X_*(u)[T_u\mathcal{N}]$ of $X : \mathcal{N} \longrightarrow T\mathcal{N}$ as a subspace of $T_u\mathcal{N}_\alpha$ then something very specific can be said about the codimension of this subspace.

THEOREM 1. Let $u \in \mathcal{M}_\alpha$ be a zero of X^α with branch points $p_1,\ldots,p_m$ of orders $r_1,\ldots,r_m$ respectively. Then the weak orthogonal complement of the image $X_*(u)[T_u\mathcal{N}]$ in $T_u\mathcal{N}_\alpha$ can be identified with the space of meromorphic vector fields on $\mathcal{D}$ tangent to S^1, and with poles at each p_i of order less than or equal to r_i. Therefore, the codimension of the range of $X_*(u)$ is the dimension of this space of meromorphic vector fields.

Using elementary methods in the theory of one complex variable one can show that the codimension is $2\Sigma r_i + 3$. Moreover theorem 1 extends to the case where there are $q_1,\ldots,q_t$ boundary branch points of (even) orders $s_1, \dots, s_t$. In this case the codimension of $X_*(u)[T_u\mathcal{N}]$ in $T_u\mathcal{N}_\alpha$ is $2\Sigma r_i + \Sigma s_i + 3$.

As a corollary we have

COROLLARY. Let $u \in \mathcal{M}_\alpha$ be a minimal surface free of branch points. Let $k \in T_u\mathcal{N}_\alpha$ be weak orthogonal to the image of $X_*(u)$. Then $k \in T_u[\mathcal{O}_G(u)]$.

We first prove Theorem 1.

Proof. (Theorem 1). Recall (from §IX) that we embedded the surface fibre bundle $\mathcal{N}$ as a submanifold of $\mathcal{A} \times H^2(S^1, \mathbb{R}^3)$ by sending a point $u_\rho \in \mathcal{N}_\rho$ to the point (ρ, u_ρ). If $\alpha \in \mathcal{A}$ a trivialization of $\mathcal{N}$ (and also of the bundles $\mathcal{S} = \bigcup \mathcal{S}_\rho$, $\mathcal{M} = \bigcup \mathcal{M}_\rho$) was given by a map $\Phi : \mathcal{W}_\alpha \times \mathcal{N}_\alpha \longrightarrow \mathcal{N} \subset \mathcal{A} \times H^2$ where $\mathcal{W}_\alpha$ is a neighborhood of α in a and Φ is defined by $\Phi(\rho,u) = (\rho, \rho(\alpha^{-1}(u)))$. Equality (13) of §IX says that for $[h_1,h_2] \in T_\alpha\mathcal{A} \times T_u\mathcal{N}_\alpha$

$$\Omega(u)\,\frac{\partial}{\partial r}\,X_*(u)\;(\Phi_*(\alpha,u)[h_1,h_2]) \tag{1}$$

$$= \left\langle \frac{d}{d\theta}\,(h_1(\alpha^{-1}(u)) + h_2),\ \frac{u_r}{\|u_r\|} \right\rangle \frac{u_\theta}{\|u_\theta\|}$$

$$+ \left\langle \frac{u_\theta}{\|u_\theta\|},\ \frac{\partial}{\partial r}\,h_1(\alpha^{-1}(u)) + \frac{\partial h_2}{\partial r} \right\rangle \frac{u_\theta}{\|u_\theta\|}\ .$$

Suppose $k \in T_u\mathcal{N}_\alpha$ is weak orthogonal to $X_*(u)[T_u\mathcal{N}]$. Then $\langle\!\langle X_*(u)h,k \rangle\!\rangle = 0$ for all $h \in T_u\mathcal{N}$. Applying equality (1), this means that

$$0 = \int_{S^1} \left\langle \frac{\partial}{\partial r}\,X_*(u)\;\Phi_*(\alpha,u)[h_1,h_2],k \right\rangle d\theta$$

for all $[h_1,h_2] \in T_\alpha\mathcal{A} \times T_u\mathcal{N}_\alpha$, or

$$(2) \qquad \int_{S^1} \left\langle k, \frac{u_\theta}{\|u_\theta\|} \right\rangle \left\langle \frac{u_\theta}{\|u_\theta\|}, \frac{\partial}{\partial r} h_1(\alpha^{-1}(u)) + \frac{\partial h_2}{\partial r} \right\rangle d\theta$$

$$+ \int_{S^1} \left\langle k, \frac{u_\theta}{\|u_\theta\|} \right\rangle \left\langle \frac{u_r}{\|u_r\|}, \frac{d}{d\theta} \{h_1(\alpha^{-1}(u)) + h_2\} \right\rangle d\theta = 0$$

Rewriting terms, integrating by parts, and using the fact that $k \in T_u \mathcal{N}_\alpha$ (2) becomes

$$(3) \qquad \int_{S^1} \left\langle \frac{\partial k}{\partial r}, h_1(\alpha^{-1}(u)) + h_2 \right\rangle d\theta$$

$$- \int_{S^1} \left\langle \frac{d}{d\theta} \{N < k, \mathcal{U} >\}, h_1(\alpha^{-1}(u)) + h_2 \right\rangle d\theta = 0$$

for all $[h_1, h_2]$, where, as before, $N = \dfrac{u_r}{\|u_r\|}$, $\mathcal{U} = \dfrac{u_\theta}{\|u_\theta\|}$.

If $u \in \mathcal{M}_\alpha$ then $h_1 \longmapsto h_1(\alpha^{-1}(u))$ has range dense in $H^{1/2}(S^1, \mathbb{R}^3)$. This implies that k satisfies the equation

$$(4) \qquad \frac{\partial k}{\partial r} = \frac{d}{d\theta} \{N(\theta) < k(\theta), \mathcal{U}(\theta) >\}$$

If we write $k(\theta) = \lambda(\theta) u_\theta$ then (4) transforms to

$$(4') \qquad \frac{\partial}{\partial r} (\lambda(\theta) u_\theta) = \frac{\partial}{\partial \theta} (\lambda(\theta) u_r)$$

These are the fundamental <u>transversality</u> <u>equations</u>. If we can show that the space of solutions of $k \in T_u \mathcal{N}_\alpha$ satisfying (4) can be "pulled back" to a meromorphic vector field on $\mathcal{D}$ with poles $p_1, \ldots, p_m$ of order less than or equal to $r_1, \ldots, r_m$ respectively, then we would be done.

Let $k \in T_u \mathcal{N}_\alpha$. Then $k(\theta) = \lambda(\theta) u_\theta = (k_1(\theta), k_2(\theta), k_3(\theta))$ where $k_j(\theta) = \lambda(\theta) \dfrac{\partial u_j}{\partial \theta}$. Define $\tilde{k}_j(\theta) = \lambda(\theta) \dfrac{\partial \tilde{u}_j}{\partial \theta}$, where $\tilde{u}_j$ is the harmonic conjugate of u_j.

LEMMA. If k satisfies the transversality equations (4) then the harmonic extension of $k + i\tilde{k}$ to the unit disc is complex analytic (each component as a map into $\mathbb{C}$ is complex analytic).

Proof (of Lemma). By the Cauchy-Riemann equations $\frac{\partial \tilde{u}_j}{\partial \theta} = \frac{\partial u_j}{\partial r}$. Therefore applying this and (4),

$$\frac{\partial \tilde{k}_j}{\partial \theta} = \frac{\partial}{\partial \theta} \left(\lambda(\theta) \frac{\partial u_j}{\partial r} \right) = \frac{\partial}{\partial r} \left(\lambda(\theta) \frac{\partial u_j}{\partial \theta} \right) = \frac{\partial k_j}{\partial r} \tag{5}$$

This implies that the harmonic extensions of k_j and $\tilde{k}_j$ satisfy the Cauchy-Riemann equations on S^1 and consequently that the harmonic extension of $k_j + i\tilde{k}_j$ is complex analytic.

Using the chain rule we have on S^1 the relation

$$k_j(\theta) + i\tilde{k}_j(\theta) = f_j'(z) \left\{ \lambda \frac{\partial x_1}{\partial \theta} + i \lambda \frac{\partial x_2}{\partial \theta} \right\}$$

where $x = (x_1, x_2)$ denotes the identity map of the unit disc to itself. Since the harmonic extensions of k_j and $\tilde{k}_j$ are analytic the quotient relations

$$\lambda \frac{\partial x_1}{\partial \theta} + i \lambda \frac{\partial x_2}{\partial \theta} = \frac{k_j(\theta) + i\, \tilde{k}_j(\theta)}{f_j'(z)}$$

$j = 1,..3$ imply that the vector field on S^1 given by $\lambda \frac{\partial x_1}{\partial \theta} + i \lambda \frac{\partial x_2}{\partial \theta}$ has an analytic extension $k_\#$ to the disc $\mathcal{D}$ minus the finite number of branch points of the minimal surface u.

From our construction it follows that if z_1 is a branch point of order r then k will have z_1 as a pole of order less than or equal to r (including the possibility that z_1 is no pole at all).

The correspondence $k \longmapsto k_\#$ sends vector fields on S^1 along u which satisfy the transversality equations to meromorphic vector fields on $\mathcal{D}$ with poles at $p_1, \ldots, p_m$ with orders less than or equal to $r_1, \ldots, r_m$ respectively. The correspondence $k \longmapsto k_\#$ is clearly bijective.

To prove the corollary, we observe that by the theorem the weak orthogonal complement of the range of $X_*(u)$ can be identified with the space of complex analytic vector fields on $\mathcal{D}$ which are tangent to S^1. Let k be in the weak orthogonal complement of $X_*(u)[T_u\mathcal{N}]$, and $k_\#$ the corresponding complex analytic vector field on $\mathcal{D}$. The vector field $k_\#$ can be integrated to a one parameter subgroup of complex analytic maps of the disc to itself. This implies that every such $k_\#$ on S^1 is of the form $\lambda(\theta)\frac{\partial x}{\partial\theta}$, $x(z) = z$ $x(z) = z$ the identity map of $\mathcal{D}$ to itself and where λ is a linear combination of $\sin\theta$, $\cos\theta$ and 1. But if $k_\#(\theta) = \lambda\frac{\partial x}{\partial\theta}$ then the corresponding $k(\theta) = \lambda\cdot\frac{\partial u}{\partial\theta}$ where λ is a linear combination of $\sin\theta$, $\cos\theta$, and 1. By the corollary to Theorem 2 §VII $k \in T_u[\mathcal{O}_G(u)]$ which completes the proof of the corollary.

Our next step will be the introduction of the notion of fine embedding and then to show that such embeddings are open in the space of all embeddings.

DEFINITION. A smooth embedding $\gamma \in A$ is <u>fine</u> if <u>all</u> minimal surfaces spanning the image $\gamma(S^1)$ are free of branch points.

This definition is due to the author, and a characterization of such embeddings is not yet known. However, Rado [26] showed that if γ is not too complicated then γ is fine. Actually Rado's original result gave conditions on γ which guarantees that any minimal surface spanning $\Gamma^\alpha = \gamma(S^1)$ is free of interior branch points. However, using Nitsches' characterization of boundary branch points (Theorem 2 §VIII) Rado's result extends to guaranteeing that any minimal surface spanning Γ^α is free of both boundary and interior branch points.

THEOREM 2 (Rado). Let $\Gamma^\gamma = \gamma(S^1)$, $\gamma \in \mathcal{A}$, have the property that there exists no point $q \in \mathbb{R}^3$ such that every plane passing through q intersects Γ^γ in at least four points. Then γ is fine.

THEOREM 3. Let $\mathcal{A} \subset H^{r+2}(S^1, \mathbb{R}^3)$ be the open subset of embeddings and let $\mathcal{F} \subset \mathcal{A}$ be the subset of fine embeddings. Then $\mathcal{F}$ is open in $\mathcal{A}$ and hence in $H^{r+2}(S^1, \mathbb{R}^3)$.

Proof. Let $\alpha \in \mathcal{F}$ and let $\mathcal{S}_\alpha$ be the codimension three submanifold of $\mathcal{N}_\alpha$ which fixes three points. By the corollary to Theorem 1 §VIII the zeros of X^α in $\mathcal{S}_\alpha \cap \mathcal{M}_\alpha$ are compact. Since harmonic functions depend linearly (and hence smoothly) on their boundary values it follows that for each zero u there is an open bounded neighborhood $\mathcal{U}$ of u and an open neighborhood $\mathcal{W}$ of α in $\mathcal{A}$ so that for $v \in \mathcal{U}$, $\rho \in \mathcal{W}$ the harmonic extension of $\rho(\alpha^{-1}(v))$ to $\mathcal{D}$ is free of branch points. Furthermore, since the set of zeros of X^α is compact, we can find an open bounded neighborhood $N \subset \mathcal{S}_\alpha \cap \mathcal{M}_\alpha$ of this set and an open neighborhood $\mathcal{W}$ of α with the properties

(i) $\bar{N} \subset \mathcal{S}_\alpha \cap \mathcal{M}_\alpha$

and (ii) for any $v \in \bar{N}$, and any $\rho \in \mathcal{W}$ the harmonic extension of $\rho(\alpha^{-1}(v))$ is free of branch points.

The fibre bundle $\pi : \mathcal{N} \longrightarrow \mathcal{A}$ of harmonic surfaces is trivializable over $\mathcal{W}$, and $\tau : \mathcal{W} \times \mathcal{N}_\alpha \longrightarrow \mathcal{N}$ given by $\tau(\rho, u) = \rho(\alpha^{-1}(u))$ is such a trivialization. Using τ, we can consider X^ρ, $\rho \in \mathcal{W}$ as a parameterized vector field $\tilde{X}^\rho$, where

$$(9) \qquad \tilde{X}^{\rho}(u) = D\tau_{\rho}^{-1} X^{\rho}(\tau_{\rho}(u))$$

and $u \longrightarrow \tau_{\rho}(u) = \tau(\rho,u)$ is, for a fixed ρ, a diffeomorphism of $\mathcal{N}_{\alpha}$ with $\mathcal{N}_{\rho}$ which when $\rho = \alpha$ is the identity map of $\mathcal{N}_{\alpha}$ to itself. This last property implies that $\tilde{X}^{\alpha} = X^{\alpha}$.

From Theorem 1 §VIII it follows that there is a neighborhood $\mathcal{W}_0 \subset \mathcal{W}$ so that for all $\rho \in \mathcal{W}_0$ the zeros of $\tilde{X}^{\rho}$ in $\mathcal{S}_{\alpha} \cap \overline{\mathcal{M}}_{\alpha}$ will have their H^2 norms uniformly bounded by some constant R. If R is large enough, the ball of radius R will contain $\overline{N}$ in its interior. Let $B_R \subset \mathcal{S}_{\alpha}$ be the intersection of this ball with $\mathcal{S}_{\alpha}$.

From Theorem 2 §X it follows that there is an $\varepsilon > 0$ and a neighborhood $\mathcal{W}_1 \subset \mathcal{W}_0$ of α with

$$\inf_{u \in B_R - N} \| \tilde{X}^{\rho}(u) \| \geq \varepsilon$$

for all $\rho \in \mathcal{W}_1$. Therefore for $\rho \in \mathcal{W}_1$ all the zeros of $\tilde{X}^{\rho}$ in $\mathcal{S}_{\alpha} \cap \overline{\mathcal{M}}_{\alpha}$ must be in $N \subset \mathcal{S}_{\alpha} \cap \mathcal{M}_{\alpha}$, and consequently any zero of X^{ρ} must be in $\rho \circ \alpha^{-1}(N)$. It immediately follows from property (ii) above that any zero of X^{ρ} in $\mathcal{S}_{\rho} \cap \overline{\mathcal{M}}_{\rho}$ is free of branch points, and the proof of Theorem 3 is complete.

Before proceeding with the proof of the open-density theorem, we need to establish an additional preliminary result.

Let $\alpha, \rho \in \mathcal{A}$ with image curves Γ^{α} and Γ^{ρ}. The map $\rho \circ \alpha^{-1} : \Gamma^{\alpha} \longrightarrow \Gamma^{\rho}$ is a diffeomorphism which we shall denote by ψ. By composition ψ induces a diffeomorphism Ψ of $\mathcal{N}_{\alpha}$ and $\mathcal{N}_{\rho}$ defined by

$$\Psi(u)(\theta) = \psi(u(\theta)) .$$

The Fréchet derivative of Ψ at a point $u \in \mathcal{N}_{\alpha}$,

$D\psi(u) : T_u\mathcal{N}_\alpha \longrightarrow T_{\Psi(u)}\mathcal{N}_\rho$ is given by the formula

(10) $$(D\Psi(u)[h](\theta) = D\psi(u(\theta))[h(\theta)]$$

where $D\psi(p) : T_p\Gamma^\alpha \longrightarrow T_{\psi(p)}\Gamma^\alpha$ denotes the finite dimensional Fréchet derivative of ψ. Both Γ^α and Γ^ρ are Riemannian submanifolds of $\mathbb{R}^3$ (their Riemannian structures are induced by that of $\mathbb{R}^3$) and as such the map $D\psi(p)$ has an adjoint

$$D\psi(p)^* : T_{\psi(p)}\Gamma^\rho \longrightarrow T_p\Gamma^\alpha$$

with respect to these structures.

The question we wish to raise is: Does $D\Psi(u)$ have an adjoint with respect to the weak Riemannian structures on $\mathcal{N}_\alpha$ and $\mathcal{N}_\rho$?

THEOREM 4. The map $D\Psi : T\mathcal{N}_\alpha \longrightarrow T\mathcal{N}_\rho$ has an adjoint $D\Psi^* : T\mathcal{N}_\rho \longrightarrow T\mathcal{N}_\alpha$ with respect to the weak Riemannian structures on $\mathcal{N}_\rho$ and $\mathcal{N}_\alpha$. Thus for any $u \in \mathcal{N}_\alpha$ and for all $h \in T_u\mathcal{N}_\alpha$, $k \in T_{\Psi(u)}\mathcal{N}_\rho$

$$\langle\!\langle D\Psi(u)[h],k\rangle\!\rangle = \langle\!\langle h,D\Psi^*(u)[k]\rangle\!\rangle .$$

In addition the dependence of $D\Psi^*$ on u is C^{r-1} smooth.

Recall that the map $L_u h = \Omega^\alpha(u)\frac{\partial}{\partial r}h$ is an isomorphism from $T_u\mathcal{N}_\alpha$ to $(T_u\mathcal{N}_\alpha)^1$, the H^1 vector fields over u. Define $D\Psi^*$ by setting $D\Psi^*(u)[k]$ equal to k_ψ, the unique solution of

(11) $$\Omega^\alpha(u(\theta))\frac{\partial}{\partial r}k_\psi(\theta) = D\psi^*(u(\theta))\Omega^\rho(\psi(u(\theta)))\frac{\partial k}{\partial r}$$

or

$$k_\psi = L_u^{-1}\{D\psi^*(u)\Omega^\rho(\psi\circ u)\frac{\partial k}{\partial r}\}$$

where $D\psi^*(p)$ is the finite dimensional Riemannian adjoint of $D\psi(p)$.

Exactly as in the proof that X^α is C^{r-1} smooth it follows that $u \longrightarrow D\Psi^*(u)$ is C^{r-1} smooth.

We are now prepared to proceed with the proof of the open density theorem.

THEOREM 5. Let $\mathcal{F} \subset \mathcal{A}$ be the open set of fine embeddings. Then there exists an open dense set $\mathcal{V} \subset \mathcal{F}$ with the property that for $\gamma \in \mathcal{V}$ all the minimal surfaces spanning γ are non-degenerate, and hence (Theorem (9) §XII) isolated.

As an immediate consequence, we obtain the

COROLLARY. An open dense set $\mathcal{V}$ of $\mathcal{F}$ has the property that every $\gamma \in \mathcal{V}$ has only a finite number of minimal surfaces spanning $\gamma(S^1)$. Thus for such γ there are at most a finite number of solutions to Plateau's problem.

Proof (of corollary). Let $\gamma \in \mathcal{V}$. We know that the minimal surfaces in $\mathcal{S}_\gamma \cap \mathcal{M}_\gamma$ are compact and all the orbits of minimal surfaces intersect $\mathcal{S}_\gamma$ in a unique point. Therefore, all minimal surfaces are represented in $\mathcal{S}_\gamma \cap \mathcal{M}_\gamma$. If the minimal surfaces spanning γ are isolated, they are isolated as points in the codimension three submanifold $\mathcal{S}_\gamma \cap \mathcal{M}_\gamma$. By compactness there can be at most finitely many such surfaces in $\mathcal{S}_\gamma \cap \mathcal{M}_\gamma$ and hence at most finitely many distinct orbits, $\mathcal{O}_G(u)$, u a zero of X^γ.

We now proceed with the proof of Theorem 5. Let $\tau : \mathcal{W} \times \mathcal{N}_\alpha \longrightarrow \mathcal{N}$ be a local trivialization of $\mathcal{N}$ with $\mathcal{W} \subset \mathcal{F}$ as neighborhood of α.

As before, for $\rho \in \mathcal{W}$, the trivialization τ induces a parameterized family of vector fields $\tilde{X}^\rho$ on $\mathcal{N}_\alpha$, given by formula (9). Let $u \longmapsto \tau_\rho(a) = \tau(\rho,u)$ be the diffeomorphism of $\mathcal{N}_\alpha$ with $\mathcal{N}_\rho$ induced by τ. Then the Fréchet derivative of τ_ρ at u is given by

$$D\tau_\rho(u) = D\Psi(u)$$

where $D\Psi(u)$ is given by formula (10), in which we drop ρ from the notation, considering it fixed. By the lemma to Theorem 8 §XII $X^\rho(w)$ is weak orthogonal to $T_w[\mathcal{O}_G(w)] \subset (T_w\mathcal{N}_\alpha)^1$.

It follows that $\tilde{X}^\rho(v)$ is weak orthogonal to the image $D\Psi^*(T_w[\mathcal{O}_G(w)])$, $w = \tau_\rho(v) = \Psi(v)$. Even though $T_w[\mathcal{O}_G(w)] \subset (T_w\mathcal{N}_\alpha)^1$ this last image makes sense because $D\Psi(v)$ extends to a linear isomorphism from $(T_v\mathcal{N}_\alpha)^1$ to $(T_{\Psi(v)}\mathcal{N}_\alpha)^1$ and it is not hard to see from formula (11) that $D\Psi^*(v)$ is an isomorphism going in the opposite direction. In addition, we have the fact that at a zero v of $\tilde{X}^\rho$, the image $\tilde{X}^\rho_*(v)[T_v\mathcal{N}_\alpha]$ is always weak orthogonal to $D\Psi^*(T_w[\mathcal{O}_G(w)])$.

It is a consequence of the defining formula (9) for $\tilde{X}^\rho$ that at a zero v of $\tilde{X}^\rho$ the kernel of $\tilde{X}^\rho_*(v)$ always contains $D\Psi^{-1}(T_w[\mathcal{O}_G(w)])$, where $w = \tau_\rho(v) = \Psi(v)$.

For each zero $u \in \mathcal{S}_\alpha \cap \mathcal{M}_\alpha$ of X^α choose connected, bounded, local exponential surfaces $\mathcal{E}$ and $\mathcal{E}'$ through u and a neighborhood N of the identity element of G so that

(a) $\bar{\mathcal{E}} \subset \mathcal{E}'$

(b) conditions (5) and (6) of Theorem 8 §XII hold for $\mathcal{E}'$

(c) $\bar{N}$ is compact

(d) $\mathcal{O}_N(\mathcal{E}')$ is an open neighborhood of u in $\mathcal{N}_\alpha$ (c.f. Theorem 6 §XII) contained in $\mathcal{M}_\alpha$. Thus $\mathcal{O}_N(\mathcal{E}') \cap \mathcal{S}_\alpha$ is also an open neighborhood of u in $\mathcal{S}_\alpha$.

For $w \in \mathcal{E}'$ let P_w be the weak orthogonal projection of $T_w\mathcal{N}_\alpha$ onto $T_w\mathcal{E}'$ and set $\tilde{P}_w = I - P_w$. Define a smooth C^{r-3} family of vector fields Y^ρ on $\mathcal{E}'$ by

$$Y^\rho(w) = P_w\tilde{X}^\rho(w)$$

Since $\mathcal{E}'$ is a finite codimensional submanifold, each Y^ρ is a Fredholm field on $\mathcal{E}'$ with respect to the connection $\mathcal{E}'$ inherits from the $\mathcal{N}_\rho$ connection induced on the ambiant manifold $\mathcal{N}_\alpha$ through the trivialization τ.

LEMMA 1. If $\mathcal{W}_0 \subset \mathcal{W}$, and $\mathcal{E}'$ are sufficiently small neighborhoods of α and u, then for $\rho \in \mathcal{W}_0$ the zeros of Y^ρ on $\mathcal{E}'$ coincide with the zeros of $\tilde{X}^\rho$ on $\mathcal{E}'$.

Proof. For $\rho = \alpha$ then Lemma 1 is precisely the result of the lemma to Theorem 8 §XII. We know from condition (b) above that for $w \in \mathcal{E}'$

$$(T_w\mathcal{N}_\alpha)^1 = (T_w\mathcal{E}')^1 \oplus T_w[\mathcal{O}_G(w)] \tag{12}$$

where

$$(T_w\mathcal{E}') \cap T_w[\mathcal{O}_G(w)] = 0 .$$

Since $D\tau_\rho$ is the identity for $\rho = \alpha$ and $\rho \longrightarrow D\tau_\rho$ is smooth, we can choose a connected neighborhood $\mathcal{W}_0 \subset \mathcal{W}$, of α and an exponential surface $\mathcal{E}'$ small enough so that for $\rho \in \mathcal{W}_0$ when replacing $T_w[\mathcal{O}_G(w)]$ in (12) by either $(D\tau_\rho)^*(T_v[\mathcal{O}_G(v)])$ or $D\tau_\rho^{-1}(T_v[\mathcal{O}_G(v)])$, $v = \tau_\rho(w)$, transversality relation (12) still holds. The remainder of the lemma now follows by the same argument used in the proof of the lemma to Theorem 8.

Consider the trivial fibre bundle $\pi_{\mathcal{E}'}: \mathcal{W}_0 \times \mathcal{E}' \longrightarrow \mathcal{W}_0$ over $\mathcal{W}_0$. The parameterized family $\{Y^\rho\}_{\rho \in \mathcal{W}_0}$ induces a vertical family Y of vector fields on the bundle $\pi_{\mathcal{E}'}$ by the formula

$$Y(\rho,w) = (\rho,w,0,Y^\rho(w))$$

when the last two coordinates belong to $T_\rho \mathcal{W}_0 \times T_w \mathcal{E}'$.

LEMMA 2. For a sufficiently small neighborhood $\mathcal{W}_1 \subset \mathcal{W}_0$ of α and a sufficiently small exponential neighborhood $\mathcal{E}'$ of u the vertical family Y is a proper Fredholm family of vector fields on $\mathcal{W}_0 \times \overline{\mathcal{E}}$.

Proof. The fact that for each $\rho \in \mathcal{W}_0$, Y^ρ is Fredholm follows (as mentioned earlier) from the fact that $\tilde{X}^\rho$ is Fredholm and $\mathcal{E}'$ is a finite codimensional submanifold of $\mathcal{N}_\alpha$.

We can assume that $\mathcal{E}'$ is open in a Hilbert space F_1 and $\mathcal{W}_0$ is open in the Hilbert space $F_0 = H^{r+2}(S^1, \mathbb{R}^3)$. It is necessary to show that there are neighborhoods $\mathcal{E}'' \subset \mathcal{E}'$ and $\mathcal{W}_1 \subset \mathcal{W}_0$ of u and α respectively with the property that for any sequence $(\rho_n,w_n) \in \mathcal{W}_1 \times \mathcal{E}''$ with $\rho_n \longrightarrow \rho \in \mathcal{W}_1$ and $Y^{\rho_n}(w_n) \longrightarrow 0$ then w_n has a convergent subsequence.

Consider the map $S : \mathcal{W}_0 \times \mathcal{E}' \longrightarrow \mathcal{W}_0 \times \mathcal{E}'$ given by $S(\rho,w) = (\rho,Y^\rho(w))$. Since Y^ρ is Fredholm S is a non-linear Fredholm map. By a theorem of Smale [29], Fredholm maps are locally proper (i.e. locally the inverse image under S of a compact set is compact). Let $\mathcal{W}_1 \times \mathcal{E}''$ be a product neighborhood of $\alpha \times u$ so that S restricted to the closure of this neighborhood enjoys the property that the inverse image of a compact set is compact. Suppose $\rho_n \longrightarrow \rho$ and $Y^{\rho_n}(w) \longrightarrow 0$ for

$(\rho_n, w_n) \in \mathcal{W}_1 \times \mathcal{E}''$. Then $S(\rho_n, w_n) \longrightarrow (\rho, 0)$. By the properness of S, $\{(\rho_n, w_n)\}$ has a subsequence which converges and hence $\{w_n\}$ has a subsequence which converges, which proves Lemma 2.

Recall that near the outset of the proof of Theorem 5 we chose connected, bounded, local exponential surfaces $\mathcal{E}$ and $\mathcal{E}'$ about each zero $u \in \mathcal{S}_\alpha \cap \mathcal{M}_\alpha$ of X^α so that conditions (a) through (d) held. To this list we now add the requirement

(e) that for each such u a neighborhood $\mathcal{W}_1$ of α and an exponential surface $\mathcal{E}'$ are chosen small enough so that Lemmas (1) and (2) are satisfied.

Recall further that $\mathcal{O}_N(\mathcal{E}') \cap \mathcal{S}_\alpha$ is an open neighborhood of u in $\mathcal{S}_\alpha$. By the compactness of the set of zeros of X^α in $\mathcal{S}_\alpha \cap \mathcal{M}_\alpha$, we can find a finite number of zeros $u_1, \ldots, u_k$ of X^α in $\mathcal{S}_\alpha \cap \mathcal{M}_\alpha$ and a connected neighborhood $\mathcal{W}_2$ of so that the corresponding exponential surfaces $\mathcal{E}_j \subset \mathcal{E}'_j$ about u_j and the neighborhood $\mathcal{W}_2$ satisfy conditions (a) through (e) and such that $\bigcup_j \mathcal{O}_N(\mathcal{E}_j)$ contains all the zeros of X^α in $\mathcal{S}_\alpha \cap \mathcal{M}_\alpha$.

From Theorem 2, §X and Theorem 1, §VIII we can deduce (c.f. Theorem 4, §XI) that there is a connected neighborhood $\mathcal{W}_3 \subset \mathcal{W}_2$ of α in $\mathcal{F}$ with the property that all the zeros of $\tilde{X}^\rho$ for $\rho \in \mathcal{W}_3$ are in $\bigcup_j \mathcal{O}_N(\mathcal{E}_j)$. Therefore, the orbit of <u>every</u> zero of $\tilde{X}^\rho$ on $\mathcal{S}_\alpha \cap \mathcal{M}_\alpha$ is represented as a zero of $\tilde{X}^\rho$ on $\bigcup \mathcal{E}_j$, and so every minimal surface in $\overline{\mathcal{M}}_\rho$, $\rho \in \mathcal{W}_3$, is represented as a zero of $\tilde{X}^\rho$ restricted to $\mathcal{E}_j$ for some j. For $\rho \in \mathcal{W}_3$ let Y^ρ_j be the corresponding projection fields on $\mathcal{E}'_j$. Therefore, by Lemma 1 and condition (e), every minimal surface in $\overline{\mathcal{M}}_\rho$, $\rho \in \mathcal{W}_3$ is represented as a zero of Y^ρ_j on $\mathcal{E}_j$ for some j. Let us place our attention on a fixed $\mathcal{E}_j$. Again consider the trivial fibre bundle $\pi_j : \mathcal{W}_3 \times \mathcal{E}'_j \longrightarrow \mathcal{W}_3$. As before, Y_j^ρ induces a vertical familv Y_j of vector fields on this bundle by the formula

$$Y_j(\rho,u) = (\rho,u,0,Y_j^\rho(u))$$

where the last two coordinates belong to $T_\rho \mathcal{W}_3 \times T_u \mathcal{E}'_j$.

There are three things we must check:

(i) Y_j is a vertical family of Fredholm vector fields of index zero

(ii) Y_j is proper on $\mathcal{W}_3 \times \overline{\mathcal{E}}_j$

(iii) Y is transverse to the zero section of the vertical subbundle of $T(\mathcal{W}_3 \times \mathcal{E}_j)$

Once we do check (i) through (iii), we may apply Theorem 2 of §I to conclude that for an open dense set $\mathcal{V}_j \subset \mathcal{W}_3$ the zeros of Y_j^ρ, $\rho \in \mathcal{V}_j$ in $\overline{\mathcal{E}}_j$ are non-degenerate. Set $\mathcal{V}_\alpha = \bigcap_j \mathcal{V}_j$. Then $\mathcal{V}_\alpha$ is open and dense in $\mathcal{W}_3$ and for each $\rho \in \mathcal{V}_\alpha$ and each j, Y_j^ρ has non-degenerate zeros on $\mathcal{E}_j$.

We have shown that for each $\alpha \in \mathcal{F}$ there is a neighborhood $\mathcal{W}_\alpha \subset \mathcal{F}$ of α and an open dense set $\mathcal{V}_\alpha \subset \mathcal{W}_\alpha$ with the property that if $\gamma \in \mathcal{V}_\alpha$ then only non-degenerate minimal surfaces span $\gamma(S^1)$. Consequently $\mathcal{V} = \bigcup_{\alpha \in} \mathcal{V}_\alpha$ is open and dense in $\mathcal{F}$ and has the property that $\gamma \in \mathcal{V}$ implies that only non-degenerate minimal surfaces span $\gamma(S^1)$. This proves Theorem 5. It remains to verify (i), (ii), and (iii).

CASE (i). That Y_j is a vertical family of Fredholm vector fields is clear from its definition. We must check that Y_j is a Fredholm family of index zero. Since $\mathcal{W}_3$ and each $\mathcal{E}_j$ are connected and the index of Fredholm vector fields is continous in the parameter ρ and constant on connected components, we need only check this at a zero $u_j \in \mathcal{E}_j$, and $\rho = \alpha$.

But

$$Y^{\alpha}_{j*}(u_j) = P_{u_j} X^{\alpha}_{*}(u_j) \mid T_{u_j}\mathcal{E}_j .$$

At u_j $T_{u_j}[\mathcal{O}_G(u_j)]$ is contained both in the kernel and cokernel of $X^{\alpha}_{*}(u_j)$. Also P_{u_j} is the weak orthogonal projection of $T_{u_j}\mathcal{N}_{\alpha} = E$ onto the weak orthogonal complement of $T_{u_j}[\mathcal{O}_G(u_j)] = E_0$ which is $E_1 = T_{u_j}\mathcal{E}_j$. We know that $E = E_0 \times E_1$ and therefore $X^{\alpha}_{*}(u_j) : E_0 \times E_1$ can be written

$$\begin{aligned} X^{\alpha}_{*}(u_j)[h_0,h_1] &= (0, P_{u_j} X^{\alpha}_{*}(u_j)[0,h_1]) \\ &= (0, Y^{\alpha}_{j*}(u_j)[h_1]) . \end{aligned}$$

It follows at once from this formula that

$\dim \operatorname{Ker} X^{\alpha}_{*}(u_j) - \dim \operatorname{Coker} X^{\alpha}_{*}(u_j) = \dim \operatorname{Ker} Y^{\alpha}_{j*}(u_j) - \dim \operatorname{Coker} Y^{\alpha}_{j*}(u_j)$.

This concludes case (i).

CASE (ii). This follows from condition (e).

CASE (iii). We must check that Y_j is transverse to the zero section of the vertical subbundle of $T(\mathcal{W}_3 \times \mathcal{E}'_j)$. For convenience let us drop the subscript j from Y. To show that Y is indeed transverse to this zero section, it is necessary to show that the total Fréchet derivative of Y, both in the parameter variable and the space variable, say $Y_*(\rho,u)$ at a zero (ρ,u) is onto $T_u\mathcal{E}'_j$.

From the corollary to Theorem 1 we know that for any zero w of X^{γ}, $\gamma \in \mathcal{F}$, the total Fréchet derivative of X maps onto the weak orthogonal complement of $T_w[\mathcal{O}_G(w)]$ in $T_w\mathcal{N}_{\gamma}$. It follows that for a zero (ρ,u), $\rho \in \mathcal{W}_3 \subset \mathcal{F}$, the total Fréchet derivative of $\tilde{X}$ at (ρ,u), $\tilde{X}_*(\rho,u)$ maps onto the weak orthogonal complement of $E_0 = (D\tau_\rho)^*(T_v[\mathcal{O}_G(v)])$ in $E = T_u\mathcal{N}_\alpha$ where $v = \tau_\rho(u)$.

For any subspace E_1 of E transverse to E_0 the projection of $\tilde{X}_*$ onto E_1 will be a surjective linear map. Since, by construction, $T_u\mathcal{E}'_j = E_1$ is such a subspace we can conclude that $P_u\tilde{X}_*(\rho,u)$ is a surjective map, where P_u is the weak orthogonal projective of $T_u\mathcal{N}_\alpha$ onto $T_u\mathcal{E}'_j$. Now

$$\tilde{X}_*(\rho,u) : H \times E \longrightarrow E$$

where $H = T_\alpha\mathcal{A}$, $E = T_u\mathcal{N}_\alpha = E_0 \times E_1 = E_2 \times E_1$, $E_2 = (D\tau_\rho)^{-1}(T_v[\mathcal{O}_G(v)])$, $v = \tau_\rho(u)$.

Therefore we can represent $\tilde{X}_*(\rho,u)$ as a map

$$\tilde{X}_*(\rho,u) : H \times E_2 \times E_1 \longrightarrow E_0 \times E_1$$

with

$$P_u\tilde{X}_*(\rho,u) : H \times E_2 \times E_1 \longrightarrow E_1$$

surjective. But

$$\tilde{X}_*(\rho,u) \mid 0 \times E_2 \times 0$$

is the zero linear map. This implies that

$$P_u\tilde{X}_*(\rho,u) \mid H \times 0 \times E_1 = Y_*(\rho,u)$$

is onto $T_u \mathcal{E}'_j$. This conclude s Theorem 5.

We end off this section with a proof of

THEOREM 6. If $\alpha \in \mathcal{V} \subset \mathcal{F}$, then there exists a neighborhood $\mathcal{V}_\alpha \subset \mathcal{V}$ of α with the property that if $\gamma \in \mathcal{V}_\alpha$ then the number of minimal surfaces spanning the image curve Γ^γ is equal to the number spanning Γ^α .

<u>Proof</u>. Since $\alpha \in \mathcal{V}$ there are only a finite number $u_1,\dots,u_r$ of non-degenerate zeros of X^α in $\mathcal{S}_\alpha \cap \mathcal{M}_\alpha$. Applying the techniques of Theorem 5, we know that we can cover the zeros of X^α by a finite number of open sets $\mathcal{O}_N(\mathcal{E}_j) \subset \mathcal{M}_\alpha$, where N is a neighborhood of the identity of G , and $\overline{\mathcal{O}_N(\mathcal{E}_i)} \cap \overline{\mathcal{O}_N(\mathcal{E}_j)} = \emptyset$ if $i \neq j$, the $\mathcal{E}_j$ being local exponential surfaces about U_j .

For $u \in \mathcal{E}_j$ define P^j_u to be the weak orthogonal projection of $T_u \mathcal{N}_\alpha$ onto $T_u \mathcal{E}_j$, and as before, set $Y^\rho_j(u) = P^j_u \tilde{X}^\rho(u)$, for $\rho \in \mathcal{W} \subset \mathcal{V}_\alpha$, $\mathcal{W}$ a neighborhood of α over which the bundle $\pi : \mathcal{N} \longrightarrow \mathcal{A}$ is trivial, and $\tilde{X}^\rho$ the parameterized family of vector fields on $\mathcal{N}_\alpha$ induced by such a trivialization. Therefore Y^ρ_j is a C^{r-3} parameterized family of vector fields on $\mathcal{E}_j$.

Since each u_j is non-degenerate $Y^\alpha_{j*}(u_j) : T_{u_j}\mathcal{E}_j \circlearrowleft$ is an isomorphism. Using this and other information, we know that if the $\mathcal{E}_j$ are taken small enough and $\mathcal{V}_\alpha \subset \mathcal{W}$ is a sufficiently small neighborhood of α then

(i) The zeros of Y^ρ_j on $\mathcal{E}_j$ are precisely the zeros of Y^ρ_j on $\mathcal{E}_j$ (Lemma 1, Theorem 5).

(ii) For $\rho \in \mathcal{V}_\alpha$ there are no zeros of $\tilde{X}^\rho$ in $\mathcal{S}_\alpha \cap \mathcal{M}_\alpha$ outside $[\bigcup_j \mathcal{O}_N(\mathcal{E}_j) \cap \mathcal{S}_\alpha] \subset \mathcal{M}_\alpha$ (this follows from Theorem 2, §X and Theorem 1, §VIII).

(iii) Y_j^ρ has a unique zero, say $v_j(\rho) \in \mathcal{E}_j$ with $v_j(\alpha) = u_j$ and the correspondence $\rho \longrightarrow v_j(p)$ is C^{r-3} (this follows from a standard application of the inverse function theorem).

From (i), (ii), and (iii) it instantly follows that $\tilde{X}^\rho$ has k-zeros $v_j(\rho)$, $j = 1,\ldots,k$, which all lie on the exponential surfaces $\mathcal{E}_j \subset \mathcal{N}_\alpha$ and where the map $\rho \longrightarrow v_j(\rho)$ C^{r-3} smooth. Consequently there are only k-zeros of X^ρ on $\mathcal{N}_\rho$. This proves Theorem 6.

COROLLARY. Let $\alpha \in \mathcal{V} \subset \mathcal{F}$. Then by definition there are a finite number of minimal surfaces $u_1,\ldots,u_k$ in $\mathcal{S}_\alpha \cap \mathcal{M}_\alpha$ spanning Γ^α . If a neighborhood $\mathcal{V}_\alpha \subset \mathcal{V}$ of α is sufficiently small there are precisely k minimal surfaces $u_1(\rho),\ldots,u_k(\rho)$ in $\mathcal{S}_\rho \cap \mathcal{M}_\rho$, spanning Γ^ρ . Moreover $\rho \longrightarrow u_j(\rho)$ are C^{r-3} functions of the embedding parameter ρ .

<u>Proof</u>. From Theorem 6 we know that if $\mathcal{V}_\alpha$ is small enough there are k-zeros $v_1(\rho),\ldots,v_k(\rho)$ of $\tilde{X}^\rho$ on $\mathcal{E}_1,\ldots,\mathcal{E}_k$ respectively with $\rho \longrightarrow v_j(\rho)$ C^{r-3} smooth. $\mathcal{S}_\rho$ was defined as the codimension three submanifold of $\mathcal{N}_\alpha$ consisting of those maps which take three points θ_j on S^1 , $j = 1,2,3$, to $\rho(\theta_j)$. Let $g_i(\rho)$ be the unique element of G which sends the three θ_j to $\alpha^{-1}(v_i(\rho)(\theta_j))$, where $v_i(\rho)(\theta_j) \in \Gamma^\alpha$ represents the map $v_i(\rho)$ evaluated at θ_j . Let $h_i(\rho)$ be the group inverse of g_i . Then for each $i = 1,\ldots,k$

$$\alpha_i(\theta_j) = v_i(\rho)(h_i(\theta_j)), \quad j = 1,2,3 .$$

Let $w_i(\rho) = v_i(\rho) \circ h_i \in \mathcal{S}_\alpha \cap \mathcal{M}_\alpha$ and let $u_i(\rho)$ be the pull back of $w_i(\rho)$ to $\mathcal{S}_\rho \cap \mathcal{M}_\rho$ by the surface bundle trivialization τ. This completes the proof of the corollary and this section.

§XIV. THE MORSE NUMBER OF MINIMAL SURFACES SPANNING A SIMPLE CLOSED CURVE AND ITS INVARIANCE UNDER ISOTOPY

In the previous section we proved the existence of an open dense set $\mathcal{V} \subset \mathcal{F}$, of the fine embeddings $\mathcal{F}$ with the property that if $\alpha \in \mathcal{V}$, then Γ^α spanned only finitely many nondegenerate minimal surfaces. We also showed (Theorem 6, §XIII) that if $\alpha \in \mathcal{V}$ then this finite number is locally constant. However, it would be nice to have a quantitative global result on the number of minimal surfaces spanning a curve.

In this final section we will show that for <u>all</u> curves $\alpha \in \mathcal{F}$ (not only those in $\mathcal{V}$) one can assign an Euler-characteristic (c.f. §I) to the vector field X^α. This is an <u>integer</u> measure of the "signed" or "Morse" number of minimal surfaces (i.e. the minimal surfaces counted with sign) spanning Γ^α. Our goal is following global result on the number of surfaces spanning the image of a fine embedding.

THEOREM. Suppose α_0 and α_1 are two fine embeddings, which are isotopic (homotopic) through fine embeddings. Then the Morse number of minimal surfaces spanning α_0 is equal to the Morse number spanning α_1.

We start, as usual, with some preliminaries.

THEOREM 1. Let $\pi_{\mathcal{F}}: \mathcal{N}_{\mathcal{F}} \to \mathcal{F}$ denote the fibre bundle of harmonic surfaces over the fine embeddings $(\pi_{\mathcal{F}} = \pi^{-1}(\mathcal{F})$; π the surface bundle projection map). Let $X: \mathcal{N}_{\mathcal{F}} \to T\mathcal{N}_{\mathcal{F}}$ denote the restriction of the C^{r-1} family of minimal surface vector fields to this bundle. Then the set of zeros $\mathcal{B}$ of X is a smooth C^{r-3} submanifold of $\mathcal{N}_{\mathcal{F}}$.

<u>Proof</u>. Since the property of being a submanifold is a local one we shall work in the neighborhood of a zero. Let $u \in \mathcal{N}_\alpha \subset \mathcal{N}$ be a zero of X^α.

Let $\mathcal{E}'$ be an exponential surface about u in $\mathcal{N}_\alpha$ and $\mathcal{W}_0$ a neighborhood of α in $\mathcal{F}$ so that conditions (a) through (d) and lemma 1 of Theorem 5 §XIII are satisfied. Recall that for $\rho \in \mathcal{W}_0$, $\tilde{X}^\rho$ is the parametrized family of vector fields on $\mathcal{N}_\alpha$ obtained by pulling X^ρ back to $\mathcal{N}_\alpha$ by a bundle trivialization τ. Y^ρ denotes the weak orthogonal projection of $\tilde{X}^\rho$ onto the exponential surface $\mathcal{E}'$. Then the zeros of the C^{r-3} family Y^ρ on $\mathcal{E}'$ are precisely the zeros of $\tilde{X}^\rho$ on $\mathcal{E}'$ (lemma 1). Let $Y:\mathcal{W}_0 \times \mathcal{E}' \to T\mathcal{E}'$ be defined by $Y(\rho,w) = Y^\rho(w)$. In the course of the proof of Theorem 5 we showed that the total Fréchet derivative of $Y, Y_*(\alpha,u)$ is surjective. By the inverse function theorem this implies that if $\mathcal{W}_1 \times \mathcal{E}, \mathcal{W}_1 \subset \mathcal{W}_0, \mathcal{E} \subset \mathcal{E}'$ is a small enough neighborhood of $(\alpha,u) \in \mathcal{W}_0 \times \mathcal{E}'$ then the set of zeros $\{(\rho,w) \in \mathcal{W}_1 \times \mathcal{E} \,|\, Y^\rho(w) = 0\} = \mathcal{B}_\mathcal{E}$ is a submanifold of $\mathcal{W}_1 \times \mathcal{E}$. Using the trivialization τ we can regard $\mathcal{W}_1 \times \mathcal{E}$ as a sunmanifold of the bundle $\mathcal{N}_\mathcal{F}$. From Theorem 6, §XII we know that $\mathcal{O}_N(\mathcal{E})$ is diffeomorphic to $N \times \mathcal{E}$, N some neighborhood of the identity of G via the map

$$(1) \qquad (g, \exp_u h) \to \exp_{g_\#(u)} A_g h$$

where exp denotes the exponential map of the weak Riemannian structure. Therefore $\mathcal{W}_1 \times \mathcal{O}_N(\mathcal{E})$ is an open neighborhood of (α,u) in $\mathcal{N}_\mathcal{F}$ diffeomorphic to the product $\mathcal{W}_1 \times N \times \mathcal{E}$. Since $\mathcal{B}_\mathcal{E}$ is a submanifold of $\mathcal{W}_1 \times \mathcal{E}$ we can apply diffeomorphism (1) to conlcude that the set of orbits $g_\#(z)$, $g \in N, z \in \mathcal{B}_\mathcal{E}$ of the submanifold of zeros $\mathcal{B}_\mathcal{E}$ of Y in $\mathcal{W}_1 \times \mathcal{E}$ is a submanifold of $\mathcal{W}_1 \times \mathcal{O}_N(\mathcal{E})$ and therefore of $\mathcal{N}$. The trivialization τ commutes with the action of G by the rule

$$\tau(\rho, g_\#(u)) = g_\#\tau(\rho,u)$$

From this equivalence and the equivariance of X under the action of G it follows that $\tilde{X}$, where $\tilde{X}(\rho,w) = \tilde{X}^{\rho}(w)$, is equivariant under G. Therefore the set of orbits $\mathcal{B}$ of elements in $\mathcal{B}_{\mathcal{E}}$ are precisely all the zeros of $\tilde{X}$ on $\mathcal{W}_1 \times \mathcal{O}_N(\mathcal{E})$. Pushing back to the bundle $\mathcal{N}_{\mathcal{J}}$ by the trivialization τ we can conclude that for every zero u there is a neighborhood $\mathcal{U}$ of u in $\mathcal{N}_{\mathcal{J}}$ with the property that the zeroes $\mathcal{B}$ of X in $\mathcal{U}$ is a submanifold of $\mathcal{U}$. This concludes Theorem 1.

REMARK 1. R. Böhme has shown that in $\mathbb{R}^3$ one can (using the Wierstrass representation theorem) actually produce coordinate charts for the manifold $\mathcal{B}$.

REMARK 2. If the action of the conformal group G on $\mathcal{N}_{\alpha}$ were smooth then the proof of Theorem 1 could be simplified considerably and would not require the machinery of sprays, exponential maps and weak projections.

Set $\mathcal{S}_{\mathcal{J}} = \bigcup_{\alpha \in \mathcal{J}} \mathcal{S}_{\alpha}$ be the subbundle of $\mathcal{N}_{\mathcal{J}}$ consisting of those maps which fix three points. $\mathcal{S}$ is a codimension three submanifold of $\mathcal{N}$. For every zero u of X, $u \in \mathcal{S}_{\alpha}$

$$T_u[\mathcal{O}_G(u)] \cap T_u\mathcal{S}_{\alpha} = 0 \tag{2}$$

If $\mathcal{B}$ again denotes the submanifold of zeroes of X in $\mathcal{N}_{\mathcal{J}}$ then $T_u[\mathcal{O}_G(u)]$ is a subspace of $T_u\mathcal{B}$. This follows from the fact that the orbit of any zero $u \in \mathcal{B}$ is also in $\mathcal{B}$. Since $\mathcal{S}_{\mathcal{J}} \subset \mathcal{N}_{\mathcal{J}}$ has codimension three in $\mathcal{N}_{\mathcal{J}}$ statement (2) implies that $\mathcal{S}$ intersects $\mathcal{B}$ transversely. The intersection $\mathcal{S}_{\mathcal{J}} \cap \mathcal{B}$ is clearly non-empty and consequently $\mathcal{S}_{\mathcal{J}} \cap \mathcal{B}$ is a submanifold of $\mathcal{N}_{\mathcal{J}}$. We state this formally in

THEOREM 2. The subbundle $\mathcal{S}_{\mathcal{J}}$ of $\mathcal{N}_{\mathcal{J}}$ is a C^r codimension three submanifold of the C^r manifold $\mathcal{N}_{\mathcal{J}}$ which intersects the C^{r-3} submanifold $\mathcal{B}$ transversely. Therefore $\mathcal{Z} = \mathcal{S}_{\mathcal{J}} \cap \mathcal{B}$ is a submanifold of $\mathcal{N}_{\mathcal{J}}$.

Let $\pi_{\mathcal{Z}} \to \mathcal{A}$ denote the restriction of $\pi_{\mathcal{A}} : \mathcal{N}_{\mathcal{A}} \to \mathcal{A}$ to the submanifold $\mathcal{Z}$. The following result is a slight modification of lemma 3 of [33].

THEOREM 3. The map $\pi_{\mathcal{Z}}$ is a C^{r-3} nonlinear proper Fredholm map of index zero (e.g. see [8] or [29]). An embedding $\alpha \in \mathcal{A}$ has the property that all minimal surfaces spanning α are nondegenerate (i.e. $\alpha \in \mathcal{V}$) iff α is a regular value of $\pi_{\mathcal{Z}}$.

REMARK. Using Theorem 3 and applying the Smale-Sard theorem [29] to the map $\pi_{\mathcal{Z}}$ gives us an alternate proof of the open density theorem 5 §XIII.

We would now like to define the <u>Morse</u> number of minimal surfaces spanning a curve $\alpha \in \mathcal{A}$.

First, if $\alpha \in \mathcal{V}$, then there are only finitely many nondegenerate minimal surfaces spanning Γ^α. Let $u_1, \ldots, u_p$ in $\mathcal{S}_\alpha$ represent these minimal surfaces. At each u_j the covariant derivative $\nabla X^\alpha(u_j)$ of X^α at u_j with respect to the connection induced by the weak Riemannian structure (c.f. §V) belongs to $\mathcal{L}_C(T_{u_j}\mathcal{N}_\alpha)$ and restricts to an element of $G\mathcal{L}_C(E)$, where $E = (T_{u_j}[\mathcal{O}_G(u_j)])^\perp$ is the weak orthogonal complement of $T_{u_j}[\mathcal{O}_G(u_j)]$, the tangent space at u_j to the orbit of u_j.

For any Banach space F the group $G\mathcal{L}_C(F)$ has two components $G\mathcal{L}_C^+(F)$ and $G\mathcal{L}_C^-(F)$. We define the sign of the covariant derivative of X^α at u_j by

$$\operatorname{sgn} \nabla X^\alpha(u_j) = \begin{cases} +1 \text{ if } & \nabla X^\alpha(u_j) \in G\mathcal{L}_C^+(E) \\ -1 \text{ if } & \nabla X^\alpha(u_j) \in G\mathcal{L}_C^-(E) \end{cases}$$

and the <u>Euler-characteristic</u> $\chi(X^\alpha)$ by

$$\chi(X^{\alpha}) = \sum_{j} \operatorname{sgn} \nabla X^{\alpha}(u_j)$$

which, by its definition, is an integer. For $\alpha \in \mathcal{V}$ we understand $\chi(X^{\alpha})$ to be the Morse number of minimal surfaces spanning Γ^{α}. We shall now define this number for a $\alpha \in \mathcal{F}$ but not necessarily in $\mathcal{V}$. This would say that for any fine embedding there is a topological measure of the number of minimal surfaces spanning any fine embedding although the actual number may be infinite.

REMARK. It is not known whether for $\alpha \in \mathcal{F}$ the actual geometric number of minimal surfaces spanning Γ^{α} can be infinite. A proof of finiteness may come from a detailed study of the bifurcation behaviour of the vector fields X^{ρ}, $\rho \in \mathcal{F}$.

Now suppose $\alpha \in \mathcal{F}$. Since $\mathcal{V} \subset \mathcal{F}$ is open and dense we can find an embedding $\alpha_1 \in \mathcal{V}$, with α_1 isotopic to α through fine embeddings.

DEFINITION. Set

(3) $$\chi(X^{\alpha}) = \chi(X^{\alpha_1})$$

We must check that this number is well defined. This depends on the next

THEOREM 4. Let $\alpha_0, \alpha_1, \in \mathcal{V}$ be two fine embeddings with α_t, $t \in [0,1]$, an isotopy between α_0 and α_1 through fine embeddings. Then $\chi(X^{\alpha_0}) = \chi(X^{\alpha_1})$.

Proof. Let $\pi_{\mathcal{Z}}: \mathcal{Z} \to \mathcal{F}$ be the projection map of Theorem 3. Then α_0 and $\mathcal{F}$ α_1 are regular values of $\pi_{\mathcal{Z}}$ and $\pi_{\mathcal{Z}}^{-1}(\alpha_i)$ are the zeros of X^{α_i} in $\mathcal{S}_{\alpha_i}$, $i = 0,1$. Since α_0 and α_1 are homotopic through fine embeddings and is open in $\mathcal{A}$ we can find a C^1 isotopy ρ_t with $t \to \rho_t$ an embedding of $[0,1]$ into $\mathcal{F}$ with $\rho_0 = \rho_0, \rho_1 = \alpha_1$. By a theorem of Smale (Theorem 3.1

[29]) there is an isotopy γ_t with $\gamma_0 = \alpha_0$, $\gamma_1 = \alpha_1$, $t \to \gamma_t$ an embedding and such that π is transverse to $t \to \gamma_t$. From this it follows that $\pi^{-1}(\gamma[0,1])$ is a one-dimensional submanifold of $\mathcal{Z}$ with boundary $\pi^{-1}(\alpha_0)$ and $\pi^{-1}(\alpha_1)$ which established a cobordism between the zeros of X^{α_0} in $\mathcal{S}_{\alpha_0}$ and those of X^{α_1} in $\mathcal{S}_{\alpha_1}$. Using the standard cobordism arguments of Pontryagin [25] as generalized to infinite dimension by Elworthy and Tromba [8], [9], we see that $\chi(X^{\alpha_0}) = \chi(X^{\alpha_1})$.

COROLLARY. Definition (3) is well defined.

We conclude this section and this paper with a proof of

THEOREM 5. Let $\gamma_0, \gamma_1 \in \mathcal{F}$ be fine embeddings with γ_o isotopic to γ_1 through fine embeddings. Then the Morse number of minimal surfaces spanning Γ^{γ_1} is equal to the Morse number spanning Γ^{γ_o}.

Proof. Choose $\alpha_0, \alpha_1 \in \mathcal{Y}$ close enough to γ_0 and γ_1 respectively so that α_i is isotopic to γ_i and α_0 is isotopic to α_1. Applying Theorem 4 and definition (3) we have

$$\chi(X^{\alpha_0}) = \chi(X^{\alpha_0}) = \chi(X^{\alpha_1}) = \chi(X^{\gamma_1})$$

REMARK 1. The Euler characteristic has an interpretation in terms of the Hessian of Dirichlet's functional E on $\mathcal{N}^{\alpha}$. Suppose $\mathcal{U}_1 \ldots \mathcal{U}_N$ are the minimal surfaces spanning the fine curve α and are non-degenerate. Then the dimension β_i of the subspace on which the Hessian $D^2E(\mathcal{U}_i)$: $T_{u_i}\mathcal{N}_\alpha \times T_{u_i}\mathcal{N}_\alpha \to \mathcal{R}$ is negative definite (excluding of course the 3-dimensional subspace which comes from the section of the conformal group is finite. Then

$$\chi(X^{\alpha}) = \sum_{i=1}^{N} (-1)^{\beta_i}.$$

REMARK 2. Reinhold Böhme [40],[41] has observed that the fine embeddings are not dense in the space of all embeddings in $\mathbb{R}^3$. However, all of the results of this paper can be extended to $\mathbb{R}^n$, $n > 3$. In a future paper Böhme and the author will show that the fine embeddings are dense in $\mathbb{R}^n$ for $n > 3$.

REMARK 3. We have already seen that at a branched minimal surface there are additional singularities at the zeroes of the minimal surface vector field. This suggested to the author a stratification of the set of minimal surfaces in the fibre bundle η. For $\mathbb{R}^3$ such a stratification has already been carried out by R. Böhme [40], although in a different setting. A fibre bundle stratification of the zeros of X^α for arbitrary codimension has been accomplished jointly by the author and Böhme.

REMARK 4. In this paper we use the space $H^2(S^1, \mathbb{R}^3)$. However we can use any manifold of maps functor for which we have elliptic regularity theory. In his Rio lectures (summer 1976) the author presented the theory in the Hölder class of functions.

Institute for Advanced Study
Princeton, New Jersey 08540
and
University of California
Santa Cruz, California

REFERENCES

[1] Agmon, S., Douglas, A., and Nirenberg., L.; "Estimates near the boundary for solutions of elliptic partial differential equations satisfying general boundary conditions I", Comm. on Pure and Appl. Math., 12, (1959), 623-727.

[2] Böhme, R., and Tomi, F.: "Zur struktur der losungsmenze des Plateau problems", Math. Z., 133(1973), 1-29.

[3] Calderon, A.P.; "Lebesque spaces of differentiable functions and distributions", AMS symposium in pure math, vol. 4.

[4] Courant, R.; "Dirichlets' Principle", Interscience, New York 1950.

[5] Douglas, J.; "Solution to the problem of Plateau"; Trans. Amer. Math. Soc. 33(1931), 263-321.

[6] Ebin, D., and Marsden, J.; "Groups of diffeomorphisms and the motion of an incompressible fluid", Ann. Math. (2) 92(1970), 102-162.

[7] Eliasson, H.; "Geometry of manifolds of maps", J. Diff. Geom. 1, (1967).

[8] Elworthy, K.D., and Tromba, A.J.; "Differential structures and Fredholm maps on Banach manifolds", Proc. Sympos. pure Math., AMS, vol. XV, 45-94.

[9] ______________________________; "Degree theory on Banach manifolds", Proc. Sympos. pure Math., AMS, vol. XVIII, 86-94.

[10] Gulliver, R., and Lesley, F.,; "On boundary branch points of minimizing surfaces", Arch. Rat. Mech. Anal., 52(1973), 20-25.

[11] Heinz, E., "Uber das randverhalten quasilinear elliptischen systeme mit isothermen parametern", Math., Z., 113(1970), 99-105.

[12] Heinz, E., and Tomi, F.;"Zu einem satz von Hilderbrandt über das randverhalten von minimalflachen", Math. Z., 111(1969), 372-386.

[13] Hilderbrandt, S.; "Boundary behavior of minimal surfaces", Arch. Rat. Mech. Anal., 35(1969), 47-81.

[14] Hormander, L.; "Linear partial differential operators", Springer, Berlin-Hieldeberg-New York, (1964).

[15] Lang, S.;"Introduction to differentiable manifolds", New York, (1962).

[16] Levy, H.; "On the boundary behavior of minimal surfaces", Proc. Nat. Acad. Sci. U.S.A., 37(1951), 103-110.

[17] Nitsche, J.C.C.; "The boundary behavior of minimal surfaces, Kellogg's theorem and branch points on the boundary", Inventiones Math., 8(1969), 313-333.

[18] ______________; "A new uniqueness theorem for minimal surfaces", Arch. Rat. Mech., 52, 319-329.

[19] Osserman, R.; "A proof of the regularity everywhere of the classical solution of Plateaus' problem", Ann. Math. 91(1970), 550-569.

[20] Palais, R.; "Morse theory on Hilbert manifolds", Top. 2, (1963), 299-340.

[21] __________; "Lusternik-Schnirelman theory on Banach manifolds", Top. 5, (1966), 115-132.

[22] __________; "On the homotopy type of certain groups of operators", Top.3, (1965), 271-279.

[23] __________; "Foundations of global non-linear analysis", Benjamin, New York, (1968).

[24] Penot, J.P.; "Une methode pour construire des varietes d'applications au moyen d'un plongement", C.R.Acad. Sc. Paris, t266, (1968), 625-627.

[25] Pontryagin, L.S.; "Smooth manfolds and their applications in homotopy theory", AMS Translations, series 2, vol. 11.

[26] Rado, T.; "On the problem of Plateau", Ergebnisse der Mathematik und Ihrer Grenzgebeite, Springer-Verlag, Berlin (1933).

[27] ________; "The problem of least area and the problem of Plateau", Math Z., 32(1930), 763-796.

[28] Smale, S.; "Morse theory and a non-linear generalization of the Dirichlet problem", Ann. Math., 80(1964), 382-396.

[29] ________; "An infinite dimensional version of Sards' theorem", Amer. J. Math., 87(1965), 861-866.

[30] Tomi, F.; "On the local uniqueness of the problem of least area", Arch. Rat. Mech. Anal., 52, 312-318.

[31] Tsuji, M.; "Potential theory in modern function theory", Tokyo, Maruzen, (1959).

[32] Tromba, A.J.; "A generalized approach to Morse theory", (to appear in J. Diff. Geom.).

[33] ___________; "Fredholm vector fields and a transversality theorem", J. Fund. Anal., Vol. 23, No. 4, December, 1976.

[34] ___________; "The Morse-Sard-Brown theorem for functionals and the problem of Plateau", (to appear in Amer. J. Math.).

[35] __________; "The Euler characteristic of vector fields on Banach manifolds and a globalization of Leray-Schander degree", (to appear).

[36] _________; "Almost Riemannian structures on Banach manifolds, the Morse lemma and the Darboux theorem", Can. J. Math., Vol. XXVIII, No. 3, pp. 640-652.

[37] _________; "Theorie de Morse sur les varietes de Banach", C.R. Acad. Sc. Paris, (1975).

[38] Alt, H. W.; Verzwergungspunkte von H-Flächen #1, Math Zeitschrift, vol. 127, 333-362.

[39] Alt, H. W.; Verzweigungspunkte von H-Flächen #2, Math Annalen, vol. 201, 33-56.

[40] Böhme, R.; Stability of minimal surfaces (to appear).

[41] Böhme, R.; Plateauprobleme für minimal flachen in $\mathbb{R}^3$, die endlich viele Lösungen haben (to appear).

[42] Hildebrandt, S., Kinderlehrer, D., and Nitsche, J. C. C.; The boundary regularity of minimal surfaces, Annali della S.N.S. XXIII (1969) 711-744.

[43] Kinderlehrer, D. and Lewy, H.; Minimal surfaces whose boundaries contain spikes, J. Math and Mech. (Indiana) 19.9 (1970) 829-853.